POINT-OF-USE/ENTRY TREATMENT OF DRINKING WATER

POINT-OF-USE/ENTRY TREATMENT OF DRINKING WATER

U.S. Environmental Protection Agency

Cincinnati, Ohio and Washington, DC

American Water Works Association

Denver, Colorado

NOYES DATA CORPORATION

Park Ridge, New Jersey, U.S.A.

Library of Congress Catalog Card Number: 90-7628
ISBN: 0-8155-1249-X
ISSN: 0090-516X
Printed in the United States

Published in the United States of America by
Noyes Data Corporation
Mill Road, Park Ridge, New Jersey 07656

10 9 8 7 6 5 4 3 2 1

Library of Congress Cataloging-in-Publication Data

Point-of-use/entry treatment of drinking water / U.S. Environmental Protection Agency and American Water Works Association.
p. cm. -- (Pollution technology review, ISSN 0090-516X ; no. 188)
Papers presented at the Conference on Point-of-Use Treatment of Drinking Water, held in Cincinnati, Ohio, Oct. 6–8, 1987.
Includes bibliographical references and index.
ISBN 0-8155-1249-X :
1. Drinking water--Purification--Congresses. 2. Household appliances--Congresses. I. United States. Environmental Protection Agency. II. American Water Works Association. III. Conference on Point-Of-Use Treatment of Drinking Water (1987 : Cincinnati, Ohio) IV. Series.
TD433.P63 1990
628'.72--dc20 90-7628
CIP

Foreword

This book provides information on the application of point-of-use (POU) and point-of-entry (POE) systems for treating drinking water. It is based on the papers presented at a recent conference held in Cincinnati, Ohio. This was the first conference of this type ever held, and it covered both administrative and technical aspects of utilizing POU/POE systems to solve individual and small community drinking water problems.

Until recent years, the principal use for POU systems was to improve the aesthetic quality (taste and odor) of drinking water provided by a municipal or private water supply. With the passage of the Safe Drinking Water Act and a greater concern by the public on health issues, consumers have begun to purchase these systems for health reasons as well as aesthetic reasons. With this increase in use of POU systems for health reasons, many issues have been raised by the regulatory agencies and water utility industry over the role of POU/POE systems in solving drinking water problems. The book presents discussions on many of these issues.

The information in the book is from *Proceedings: Conference on Point-of-Use Treatment of Drinking Water,* co-sponsored by the U.S. Environmental Protection Agency and the American Water Works Association, issued by the U.S. Environmental Protection Agency, June 1988.

The table of contents is organized in such a way as to serve as a subject index and provides easy access to the information contained in the book.

Advanced composition and production methods developed by Noyes Data Corporation are employed to bring this durably bound book to you in a minimum of time. Special techniques are used to close the gap between "manuscript" and "completed book." In order to keep the price of the book to a reasonable level, it has been partially reproduced by photo-offset directly from the original report and the cost saving passed on to the reader. Due to this method of publishing, certain portions of the book may be less legible than desired.

ACKNOWLEDGMENTS

Numerous individuals were responsible for the success of this conference. Planning of the program was accomplished by Tom Sorg, Kim Fox, Frank Bell, and James Smith of the U.S. EPA and Jon DeBoer of the AWWA. A special thanks is given to Patricia Cooke who arranged for the conference facilities and to all the U.S. EPA employees who helped during registration and provided other support throughout the conference. Sincere appreciation is extended to Lisa Moore, Carolyn McGill, and other members of the JACA Corporation who were responsible for preregistration and preparation of the Proceedings, and who also provided help throughout the entire conference. Last, but not least, all the speakers and moderators are acknowledged for their excellent presentations and the preparation of their written pages.

NOTICE

The materials in this book were prepared as accounts of work sponsored by the U.S. Environmental Protection Agency and the American Water Works Association. On this basis the Publisher assumes no responsibility nor liability for errors or any consequences arising from the use of the information contained herein.

The following chapters have been reviewed in accordance with the U.S. Environmental Protection Agency's peer and administrative review policies and approved for presentation and publication:

- Point-of-Entry and Point-of-Use Devices for Meeting Drinking Water Standards
- Regulatory Requirements for Point-of-Use Systems
- Microbiological Studies of Granular Activated Carbon Point-of-Use Systems
- Health Studies of Aerobic Heterotrophic Bacteria Colonizing Granular Activated Carbon Systems
- Community Demonstration of POU Systems Removal of Arsenic and Fluoride
- POU/POE Point-of-View (Discussion by F. Bell)

The remaining chapters were not prepared with U.S. EPA financial support and the contents do not necessarily reflect the views of the Agency; therefore, no official endorsement should be inferred.

Mention of trade names or commercial products does not constitute endorsement or recommendation for use by the Agency or the Publisher. Final determination of the suitability of any information or procedure for use contemplated by any user, and the manner of that use, is the sole responsibility of the user. The reader is warned that caution must always be exercised when dealing with water treatment materials or processes which might be hazardous, and expert advice should be sought at all times.

All information pertaining to law and regulations is provided for background only. The reader must contact the appropriate legal sources and regulatory authorities for up-to-date regulatory requirements, and their interpretation and implementation.

Contents and Subject Index

OVERVIEW OF POINT-OF-USE AND POINT-OF-ENTRY SYSTEMS

Lee T. Rozelle
Olin Corporation
Cheshire, CT 06410

The 1986 amendments to the Safe Drinking Water Act require that 83 contaminants must be regulated within three years of signing the act, that is by June, 1989. EPA has set up a schedule to comply with the amendments including the eight volatile organic chemicals currently finalized.

The greatest regulatory burden will be on the 38,000 community systems serving less than 500 people. When violations are incurred, modification or installation of conventional water treatment systems may be too costly and these communities may apply for variances or exemptions. Although unregulated by the Safe Drinking Water Act, the 850,000 rural systems with two to 14 connections and an additional 9,000,000 individual rural systems, will certainly be affected by this act.

The utilization of proven water treatment technologies at the point-of-use/point-of-entry (POU/POE) offers a potentially viable and cost effective method of reduction of chemical contaminants to acceptable levels in drinking water. In fact, in situations where contaminants pose an unreasonable risk to health, the option of point-of-use or bottled water is being proposed by U.S. EPA as a condition for receiving a variance or exemption. This would be on a temporary basis until compliance with the regulations is achieved. Also there are certain conditions for use such as certification, bacterial safety, etc.

Point-of-entry is acceptable to EPA for long term use in contaminant removal from drinking water supplies, although it has not been given Best Available Technology status. Again there are certain conditions for use, similar to those of point-of-use.

A point-of-use treatment device consists of equipment applied to selected taps used for the purpose of reducing contaminants in water at each tap.

A point-of-entry treatment device consists of equipment applied to water entering a house or building for the purpose of reducing contaminants distributed throughout the house or building.

Point-of-use systems are commonly placed in the following locations at the sink:

- *Counter top.* A counter top device normally fits through a connection to the faucet on the sink and rests on the counter or in the sink.
- *Faucet Mounted.* A faucet mounted filter is attached directly to the end of the faucet.
- *Under Sink Cold Tap.* This device fits onto the cold water line and treats all the cold water that flows through the faucet.
- *Under the Sink Line Bypass.* This device taps onto the cold water line, and after flowing through the lines to a reservoir (in some cases), exits through a special spigot attached to the sink.

Point-of-entry systems are placed where the household water enters the house (but normally after the outside outlets).

Common POU/POE technologies and their placement are shown in Table 1. Ultraviolet radiation is also an effective POU/POE technology for reduction of microorganisms. Placement could include counter top and under the sink as well as point-of-entry.

Table 1. Common POU/POE Technologies and Their Placement

Technology	Normal Placement
Particulate Filters	All POU Placements POE
Adsorption Filters	All POU Placements POE
Reverse Osmosis	Countertop Undersink Line Bypass
Ion Exchange	POE Potentially All POU Placements
Distillation	Countertop

Particulate filters at the point of use normally have 3 to 60 µm (0.0001 to 0.002 in) ratings. These filters consist of media such as spun bonded materials, foam (molded in place), wound string, fabric, membranes, and granular activated carbon (GAC). The point-of-use particulate filters are typically 25.4

cm (10 in) in height and 7.6 cm (3 in) in diameter. Point-of-entry particulate filters resemble household softeners with media consisting of sand, granite, anthracite, etc.

The most common adsorption filter for contaminant reduction in drinking water is granular activated carbon (GAC). These filters reduce common tastes and odors, some turbidity, residual chlorine, radon, and many organic contaminants with varying degrees of efficiency based on molecular structure and equipment design. The most common design is a cartridge containing a loose carbon bed. This bed-type filter could also contain activated alumina as the media for fluoride and arsenic (V) reduction. Fused carbon filters and precoat filters also are utilized. Precoat filters usually consist of powdered activated carbon and/or diatomaceous earth applied to the influent side of the filter.

Reverse osmosis (RO) is considered the "high tech" method for reduction of dissolved solids. Currently it is more applicable at the point-of-use than point-of-entry. Typical RO membranes remove total dissolved solids (TDS) with such efficiency that the treated water may become aggressive and dissolve metals from the water pipes; blending may be necessary. Also, at the point entry for RO there would be a need for a holding tank and a recirculation pump resulting in a more expensive treatment system.

The reverse osmosis systems normally consist of a particulate filter followed by an optional activated carbon filter (if a chlorine sensitive membrane is used), an RO module, a water reservoir containing a pressurized rubber bladder (approximately 9.5 l [2-1/2 gal] capacity), a final activated carbon filter (to remove any residual taste and odor), and the special spigot on the sink. Household under-the-sink units operate efficiently at pressures between 2.8 and 4.9 kg/cm^2 (40 and 70 psi) on nonbrackish raw waters with up to 2,000 mg/l of TDS. The flow rate through the spigot is typically between 0.03 and 0.06 l/s (0.5 and 1 gpm).

Cation exchange has been used for water softening for over 50 years. However ion exchange can apply to selective inorganic contaminant removal using either cation or anion exchange resins.

Distillation has historically been known to be effective and has been utilized for producing contaminant free water.

The maintenance of point-of-use and point-of-entry systems is necessary to maintain their effectiveness for contaminant removal. The following summarizes maintenance procedures of current point-of-use/entry devices utilizing proven technologies:

- *Particulate filters.* Particulate filters at the point-of-use are replaced before clogging, when slow flow is observed. At the point-of-entry these filters are backwashed periodically.
- *Granular activated carbon.* Granular activated carbon filters must be replaced before breakthrough of the contaminants. Many units contain a shutoff or alarm device to indicate when a certain volume of water has been filtered. If no shutoff device is on the unit, the filter either periodically replaced by a qualified dealer or, if listed by the National Sanitation Foundation (NSF) Standard 53, a 100 percent safety factor is used for replacement based on volume flow. That is, if a 3,790-l (1,000-gal) capacity is claimed, it must be effective for 7,580-l (2,000-gal) to pass NSF Standard 63.
- *Point-of-use RO systems* require a periodic replacement of filters and the RO modules (to avoid loss of efficiency due to membrane fouling or deterioration). According to NSF Standard 58, the RO module must be replaced when the conductivity rejection is below 75 percent or at a value necessary to maintain drinking water compliant with the Safe Drinking Water Act. With proper maintenance of the prefilters, the cellulose acetate modules are normally replaced after one and a half to two years of service and the polyamide modules replaced between two and four years of service. The GAC and particulate filters are normally replaced every six to 12 months.
- *Ion Exchange.* Ion exchange units are normally regenerated with sodium chloride.
- *Distillation.* Distillers must be cleaned due to scaling.
- *Activated Aluminia.* Activated aluminia is regenerated by sodium hydroxide and then acidified for adsorption.

The costs of point-of-use/point-of-entry units and their replacement follow:

- *Particulate filters* typically cost between $20 and $100.
- *Granular activated carbon* filters typically cost between $50 and $300 with the lowest replacement cost about $20. For the point-of-entry the cost range of granular activated carbon filters is $800 to $l,000, with $200 to $400 replacement costs.
- *Reverse osmosis* devices vary in price. The counter top devices range from $100 to $300. Under-the-sink RO devices containing cellulose acetate membranes range from $300 to $600 with replacement cost at $50 to $60 for the CA membrane element. Under-the-sink RO devices containing thin film composite membranes range from $400 to $800 with replacement cost of the membrane element at around $100.

- *Ion exchange* costs vary from $300 to $2,000 depending on the resin (anion exchange resins cost more than cation exchange resins) and type of equipment.

- *Distillation* typically costs between $200 and $600.

- *Ultraviolet units* typically cost between $300 and $700.

Reverse osmosis is the best technology for reduction of inorganic contaminants considering no energy input. It may be indispensable for lead and copper, which may contaminate water from household pipes. It should be remembered, however, that reverse osmosis efficiency depends on the type of membrane used. Cellulose acetate membranes normally do not reject some contaminants as effectively as the newer thin film composite membranes. For example, nitrates are rejected up to 65 percent by cellulose acetate membranes, but up to 94 percent by thin film composite membranes under point-of-use conditions.

Selective reduction of inorganic contaminants can be carried out by cation exchange resins for ions such as radium and barium and by anion exchange resins for nitrates and arsenates.

Activated alumina is effective for reduction of fluoride and arsenic. Granular activated carbon is very effective for reduction of radon. With proper design virtually 100 percent reduction has been observed. Disposal and shielding remain an issue for general use. Distillation is also effective for inorganic removal.

Granular activated carbon is known to be the most effective and inexpensive method for removal of organic contaminants in drinking water. It must be remembered, however, that it is not perfect and the adsorption capacity for various contaminants varies. Thus, it is important to know which organic contaminants are present and their adsorption capacities for effective maintenance.

Reverse osmosis is not known for effective reduction of volatile organic chemical or low molecular weight organic contaminants. Reduction efficiency varies based on molecular weight, charge, size, shape, and relationship with the chemistry of the membrane. There are indications, however, that total organic carbon is more consistently reduced (80 to 90 percent) by reverse osmosis when compared to granular activated carbon.

When GAC is used with RO, as in many line bypass RO systems, organic reductions increase, specifically for low molecular weight organics including VOCs. This combination can result in a very effective point-of-use device for removal of contaminants.

Several field studies have been carried out using point-of-use/point-of-entry. In Suffolk County, Long Island 3,000 GAC units have been used for over four years to treat water with an average aldicarb concentration of 87 μg/l. Based on this experience, at a 100 μg/l influent concentration using 0.028 m^3 (1 cu ft] of GAC, the GAC filter life was calculated to be 170,325 l (45,000 gal) before breakthrough of 7 μg/l aldicarb.

In Rockaway Township, New Jersey 12 GAC units were used to remove concentrations from water above 100 μg/l of TCE and 1,1,1-trichloroethane. After 24 months of testing, no significant concentrations were observed in the effluent (less than 1 μg/l [1 ppb]). In Silverdale, Pennsylvania 47 GAC devices were tested using five models. With influent concentrations of TCE above 100 μg/l (100 ppb), no significant concentrations were observed in the effluent after 14 months of operation.

In Emmington, Illinois 63 reverse osmosis devices were tested for removal of fluoride and high total dissolved solids. In this one year test, the fluoride was reduced by 86 percent (from raw water concentration of 4.5 mg/l [4.5 ppm]) and the TDS reduced 79 percent (from a raw water concentration of 2,620 mg/l).

The actual costs of the point-of-use purchase, operation, and maintenance in Rockaway and Silverdale varied from $5.98 a month in Silverdale to $4.23 a month in Rockaway. An estimate of an additional $1.23 per month was made for administrative costs if used in a community of 650 customers.

In Emmington, Illinois the actual point-of-use reverse osmosis costs were $12.48 a month. It was estimated that if reverse osmosis was used for central treatment, the cost to the homeowner would be $28.50 a month.

As a comparison, bottled water used in a family of 2.8 people at 3.8 l per day (1 gpd) per person at a price of $0.22 per liter ($0.85 per gal) would cost $67.00 per month.

The following conclusions result from the study:

- Use of proven technologies at the point-of-use and point-of-entry is effective for reducing contaminants from drinking water supplies.

- Reverse osmosis and distillation are most universally effective for inorganic reduction.

- Granular activated carbon is most universally effective for organic contaminant reduction.

- Costs for small communities appear to be attractive particularly if these devices can be leased to the community avoiding up front costs.

Point-of-Entry and Point-of-Use Devices for Meeting Drinking Water Standards

Stephen W. Clark
U.S. Environmental Protection Agency
Washington, DC 20460

BACKGROUND

The Safe Drinking Water Act Amendments of 1986 require the Environmental Protection Agency (EPA) to set standards for 83 contaminants by June of 1989. These contaminants include inorganic chemicals, radionuclides, and organic chemicals. The microbial contaminants listed by Congress will be regulated by requiring filtration as a treatment technique for surface waters, and disinfection of all waters. The chemical contaminants will be regulated by setting maximum contaminant levels (MCLs). MCLs generally apply at the tap and represent an achievable, safe level of contaminants in public water supplies.

The public water supplies regulated under the Safe Drinking Water Act include all systems serving at least 25 people or 15 service connections. The EPA has by regulation created three subcategories of public water systems. Community water systems service fifteen or more connections. There are approximately 65,000 community water systems in the U.S. They range in size from very small communities to large cities like New York and Chicago. The other major category is noncommunity water systems, which serve at least 25 persons. There are over 200,000 noncommunity water systems, a category which includes restaurants, parks, factories, and other places frequented by the public. Nontransient, noncommunity water systems such as schools or workplaces serve the same 25 or more people at least six months of the year. There are approximately 20,000 nontransient, noncommunity water systems that will have to meet the same standards as community water systems. The remaining noncommunity water systems will have to meet standards for microbial contaminants and some acutely toxic chemicals like nitrate. The reason for the difference is that some toxicants (e.g., fluoride) require lifetime exposure to increase risk of diseases, whereas acute toxicants can theoretically cause diseases like hepatitis after one drink of contaminated water. The majority of all kinds of water systems are small, that is they serve less than 3,300 people or 600 service connections. Compliance is good among large, metropolitan systems, but small systems have historically lacked the money and the technical skill to operate complex water treatment plants.

Recognizing the difficulty that small systems would have complying with the many new drinking water standards, EPA considered allowing a variety of decentralized approaches. These approaches included point-of-entry devices, point-of-use devices, and bottled water.

Public comment was first sought on these decentralized approaches in the Federal Register of November 1985 (1). This notice proposed MCLs for eight volatile organic chemicals as well as criteria for the use of decentralized approaches in public drinking water systems.

MAJOR ISSUES

Although the Federal Register notice sought comment on these along with other issues, the U.S. EPA decided to conduct a public hearing on decentralized approaches in June of 1986. The three major issues discussed at this meeting were;

- Should point-of-entry (POE) devices treating all the water entering buildings connected to a public water system be considered a suitable means of compliance?
- Should point-of-use (POU) and bottled water in addition to POE be considered suitable means of compliance?
- Should POE, POU, or bottled water be considered Best Available Technology (BAT) for small systems (less than 600 persons)?

Three options were presented at this meeting. They are discussed below.

Option 1
Consider POE to be an acceptable means of compliance.

Explanation
Allow the application of POE devices to treat all the water in every building for compliance purposes.

Bottled water and POU could be considered as interim means of reducing excessive risks during emergency situations. However, POE would not be considered BAT under this option. EPA was leaning toward this option, at that time.

Discussion

1) From a human exposure standpoint POE could be considered equivalent to central treatment.

2) POE treatment methods are similar to the central treatment options that would be used for small systems.

3) From a practical perspective there are some differences:

   a) Monitoring would have to be increased to assure that each device is functioning properly (i.e., producing water meeting the MCL).

   b) Operation and maintenance is much more difficult than for a central treatment system.

   c) POE is less likely to be suitable for compliance with the microbiological standards because:

      - Microbiological safety is assured by maintaining good source waters, the application of filtration and disinfection technologies as appropriate (including the maintenance of a disinfectant residual throughout the distribution system), and maintenance of the integrity of the distribution system.

      - Protection from acutely hazardous contaminants (such as microbes) is critical, and more difficult to assure in a decentralized operation (which would naturally have less supervisory control).

      - POE devices might not be able to provide protection equivalent to central treatment because of these considerations.

   d) Compliance for some contaminants would be determined by multiple in-building samples for the POE mode.

   e) There might not be any cost advantage for the POE option over for central treatment, especially as the hydraulic capacity of the system increases.

   f) Tradition in the industry and some of the legislative history of the Safe Drinking Water Act suggest that the trend toward regionalization versus decentralization.

   g) Compatibility of POE devices with the central treatment technologies currently in place or required in the future needs to be considered.

      - Without post-disinfection, GAC adsorption POE devices would contribute microorganisms to the water supply (as with POU).

      - In addition, this could result in exposure to microbes via inhalation as well as by drinking.

4) Because POE would not be considered to be BAT, EPA would not require its installation before a variance could be granted to a water system. If a system could install POU to gain near term benefit, it would be allowed to do so. However, if the system desired a variance, it would have to install central treatment (BAT) to fulfill the statutory conditions for variances.

5) Concern was expressed that persons may still consume water from untreated taps of systems are allowed to use bottled water or POU devices for long-term compliance purposes. This is one reason why EPA was leaning toward requiring that all water provided to the consumer be treated.

Conditions for Choosing Option 1

1) Public water systems would have to maintain control and responsibility for the operation and maintenance of the POE devices.

2) A monitoring and maintenance program that assures protection of all consumers equal to that provided by the central treatment option.

3) Effective technology must be properly applied including provisions for microbiological safety.

4) All consumers in every building must be protected (i.e., have a device installed, maintained, and adequately monitored by the responsible party).

5) The POE mode of compliance must provide protection equivalent to that provided by central water treatment.

Option 2

Allow POE, POU, and bottled water as acceptable means of compliance.

Explanation

Allow POU and bottled water in addition to POE as suitable means of compliance under defined circumstances and criteria. None of these would be considered BAT under this option.

Discussion

1) Since respiratory and dermal exposure have been identified as concerns for volatile chemicals and microbiological contaminants, then all but central treatment or POE would be ruled out for these substances. Under certain circumstances POE

might not be acceptable for biological contaminants (see Option 1 discussion).

2) To allow both bottled water and POU:

a) Bottled water, meeting all standards, should be delivered to the consumers in order to be similar to POU.

b) A special monitoring scheme for bottled water would need to be developed.

c) Because POE would not be considered to be BAT, EPA would not require its installation before a variance could be granted to a water system. If a system could install POU to gain near-term benefits, it would be allowed to do so. However, if the system desired a variance, it would have to install central treatment (BAT) to fulfill the statutory conditions for variances.

Conditions for Choosing Option 2

1) The public water system would have to maintain control and responsibility for the operation and maintenance of the POU or quality control over the contents and delivery of the bottled water.

2) A special monitoring program that assures protection of all consumers equal to that provided by the central treatment would be required. It could consist of application of the Part 141 monitoring requirements.

3) Effective technology must be properly applied including provisions for microbiological safety - bottled water must meet the microbiological safety standards, too.

4) All consumers in every building must be protected (i.e., have a POU or bottled water device installed, maintained, and adequately monitored by the responsible party).

5) Bottled water is not "piped water" for human consumption, thus arguably excluding these systems from the definition of a public water system. To allow bottled water for drinking water systems EPA would therefore, have to determine that provision of bottled water, under certain conditions, is equivalent to provision of "piped water" by a public water system.

Option 3

Consider POE, POU, and bottled water to be Best Available Technology (BAT) for small systems (less than 500 persons).

Explanation

On a compound-by-compound basis, criteria would be set by which small systems could be required to use POE, POU, or bottled water in lieu of central treatment prior to being granted a variance. That is, for purposes of receiving variances to specific MCLs, POE, POU, or bottled water would have to be installed by small systems.

Discussion/Conditions for Choosing Option 3

1) The criteria used to determine BAT for decentralized treatment would differ from that for the central treatment option, and may vary by contaminant.

2) As an example, consider criteria for the determination of BAT for fluoride. The POU, POE, or bottled water must be:

a) Commercially available, and capable of satisfactorily removing fluoride from drinking water.

b) Affordable by large metropolitan public water systems.

c) "Best" based upon the following factors:

- Wide applicability,
- High cost efficiency,
- High degree of compatibility with other water treatments in use or needed for the system, and
- The ability to achieve compliance for all water in a public water system.

3) Affordability criteria would be different for large and small systems.

4) Central treatment would still be available.

5) The amount of space required for installation of POE could limit the applicability of POE throughout a system, and hence, its designation as BAT.

6) The high degree of compatibility criterion would be considered on a compound- and technology-specific basis.

7) EPA would need evidence that costs are reasonable.

SUMMARY OF EPA DECISION

After considering all public comments and through a variety of discussions at all levels of management, EPA decided that point-of-entry devices were suitable for compliance, but they were not BAT. It was also decided that POU and bottled water could be used as interim measures, but were not to be considered BAT or a means of compliance.

The decentralized approaches cannot be considered BAT because of difficulties associated with monitoring compliance and assuring effective treatment performance in a manner comparable to central treatment. Most of the public comments received by EPA were against considering the decentralized approaches BAT. The commenters cited difficulties in controlling installation, maintenance, operation, repair, and potential exposure through untreated taps. However, other commenters felt that decentralized technologies were BAT for very small systems, as these methods were often more cost effective for some small systems than central treatment.

In the final rule, POE and POU were not designated as BAT because: 1) it is more difficult to monitor the reliability of treatment performance and to control POU and POE than for central treatment; 2) these devices are generally not affordable by large metropolitan water systems; and 3) in the case of POU, not all the water is treated. In addition, POU and bottled water are not considered acceptable means of compliance with MCLs. Neither these devices nor bottled water treat all the water in the home and could result in health risks due to exposure to untreated water. Consequently, POU and bottled water are only considered acceptable for use as interim measures. That is, they may be required by a state primacy agent as a condition of obtaining a variance or exemption, if necessary to avoid an unreasonable risk to health before full compliance could be achieved. Under this rule, however, POE devices are acceptable means of compliance, because POE provides drinking water meeting standards at all taps in the house. Furthermore, these devices might be cost effective for small public water systems or nontransient, noncommunity water systems. The ultimate goal for the EPA drinking water program is to have water meeting all standards through regionalization, pure source water, or central water treatment.

FINAL REGULATIONS ALLOWING POE FOR COMPLIANCE

Introduction

The EPA promulgated a final rule in July 1987 that allowed POE devices as a means of compliance with the final MCLs for volatile organic chemicals. POU or bottled water could only be used to alleviate unreasonable risks to health during a variance or exemption period - that is, while the public water systems were attempting to come into full compliance with the MCL using POE or central treatment. A more detailed discussion of the criteria for the use of POE as a means of compliance follows.

Definitions

The final rule (Code of Federal Regulations, Part 141.2) defines POU and POE. These definitions are worth repeating here for clarity (2):

"'Point-of-entry treatment device' is a treatment device applied to drinking water entering a house or building for the purpose of reducing contaminants in the drinking water distributed throughout the house or building."

"'Point-of-use treatment device' is a treatment device applied to a single tap used for the purpose of reducing contaminants in drinking water at that one tap."

Criteria and Procedures

EPA is required to establish conditions for treatment and control of public water supplies that assure protection of public health. Specifically, EPA's primary drinking water regulations are to contain criteria and procedures to assure a supply of drinking water that dependably complies with MCLs, including quality control and testing procedures to insure compliance with such levels and to insure proper operation and maintenance of the system. It is under this authority that EPA promulgated criteria and procedures allowing the use of POE for compliance with the volatile organic chemical MCLs. As was mentioned earlier, EPA feels that philosophically the Safe Drinking Water Act, including the legislative history, emphasizes the goal of providing pure water throughout a centrally controlled facility. Realizing that this philosophical goal is not always attainable, especially for small systems that lack the necessary financial and technical resources, EPA is seeking to allow innovation in order to gain increased compliance among these systems. Historically, the largest number of violations have occurred among the very small systems. It is hoped that this rule will allow them a more accessible means of compliance, and would, in turn, increase their compliance rate. The rule specifies criteria and procedures that will hopefully assure quality and safety when POE is applied for compliance purposes.

The five criteria necessary for compliance using POE are summarized below.

Central Control

Originally, in the November 13, 1985 rule, central ownership and control were required. It would be the responsibility of the public water system (PWS) to own, operate, and maintain all parts of the treatment system. This appeared appropriate and necessary to ensure adequate control of the treatment devices so that they were working properly.

Public comments noted that while central control and responsibility were necessary, ownership of the devices was not. The PWS, while maintaining responsibility and control, could lease the treatment devices and also possibly have them operated and maintained by a service company. The major concern of EPA was that the property owners would not individually become responsible for these devices.

The final rule requires the public water system to be responsible for operating and maintaining all parts of the treatment system including each POE device. Central ownership is not necessary so long as the public water system maintains control of the operation and maintenance of the device. This includes being responsible for and supervising any service contractor acting on behalf of the public water system. Central control is appropriate and necessary to ensure that the treatment device is always functioning properly.

Effective Monitoring

The public water system must develop an effective monitoring plan and obtain state primacy agent approval before POE devices are installed for compliance with drinking water standards. POE devices are fundamentally different from central treatment in that many more devices are installed at different locations. All mechanical devices have some theoretical or empirical failure rate. As the numbers of devices applied at one public water system increase so does the probability of some devices failing. Typically, at the central water plant, the operator makes observations and measurements. This clearly becomes more difficult with POE devices since they are numerous and are generally located on private property.

A monitoring program would include some proportion of the devices, for example, 10 percent. Monitoring might rotate throughout the population on a quarterly basis. In some cases, physical inspections and flow measurements could be made on the entire population, with the proportionate sampling being done for more expensive chemical analyses. The cost of analyses for volatile organic chemicals is so high and probability of failure in a well-designed, well-operated treatment device so low that a small number of samples should be adequate. The details of this requirement remain to be determined by the state. The state and the utility should work together to formulate an adequate, yet affordable program.

Application of Effective Technology

Design review of plans and specifications for modifications or additions to water works are generally required under state authority. Almost every state requires reviews and permits for this kind of activity. EPA recognized this and mandated a similar kind of review. This review should include certification that the device will perform adequately to protect public health in the individual application contemplated. Most states are likely, and it is certainly appropriate, to accept recognized third-party certification of these devices. Third-party certification should not preempt a review to assure that the application of certified device is appropriate. There are, then, two responsibilities: 1) certification of the device for various contaminant removal situations, and 2) review to assure that the device is being applied to an appropriate situation.

Maintenance of Microbial Safety

The design and application of POE devices must consider the tendency for increases in bacterial concentrations in water treated with granular activated carbon and possibly some other technologies. At a central treatment plant, provisions can be made for granular activated carbon adsorbers to be backwashed and post-disinfection is generally practiced. In a POE situation, the disinfectant, if present, is in the incoming water. GAC is an effective media for removal of chlorine from water. It also has been shown to provide a surface for the attachment and growth of heterotrophic bacteria. Heterotrophic bacteria are not usually harmful to health, but do present two concerns. The first is that they may infect people who are sickly and have a low resistance to bacterial infection, especially of the respiratory tract. Secondly, high concentrations of heterotrophic bacteria (greater than 500 per ml), can interfere with the examination of the water for coliform bacteria. The state might require additional monitoring for heterotrophic bacteria to evaluate for possible interference with the required coliform bacterial tests. If interference is suspected or counts are high enough to be of concern via respiratory exposure, then the state might require post disinfection.

Post disinfection after a GAC adsorption unit would require an ultraviolet device, or a chlorinator with an adequate contact tank. The contact tank and chlorinator can be designed in accordance with standard procedures used for providing disinfection of single buildings, using noncommunity water wells. Post disinfection would clearly increase the cost of POU treatment and might bias the economics toward central treatment.

Protection of All Consumers

Every building connected to a public water system must have a POE device installed, maintained, and adequately monitored. The device should provide treated water to every potable water tap within each building. It is up to the state to determine if some taps within or outside certain buildings may remain untreated. For example, if the state allows nonpotable water to be used for car washing, then this portion of the water can be untreated. Other nonpotable uses might include lawn watering devices, aesthetic fountains, industrial cooling, and fire protection. As previously mentioned there is concern, especially with volatile chemicals, for respiratory exposure. Therefore, the definition of nonpotable water should never extend to living or nonindustrial working spaces where these kinds of exposure are possible.

SUMMARY

The goal of these five criteria is to assure that when POE devices are applied by public water systems for compliance with drinking water standards, the water is as safe as the tlme-tested methods of central water treatment. These criteria provide for adequate public health protection, while at the same time allowing for an innovative, decentralized approach (i.e., POE). Hopefully, this approach will allow a cost effective means of compliance for small systems that have had the most violations of EPA's drinking water standards. The criteria developed by EPA will be adopted and implemented by the states with the goal of providing safe drinking water to all communities.

REFERENCES

1. National Primary Drinking Water Regulations, Volatile Synthetic Organic Chemicals, Final Rule and Proposed Rule. Fed. Reg. 50:219:46880-46933. November 13, 1985.

2. National Primary Drinking Water Regulations, Synthetic Organic Chemicals, Monitoring for Unregulated Contaminants, Final Rule. Fed. Reg. 52:130:25690-25717. July 8, 1987.

REGULATORY REQUIREMENTS FOR POINT-OF-USE SYSTEMS

Ruth Douglas
Registration Division
Office of Pesticides and Toxic Substances
U.S. Environmental Protection Agency
Washington, DC

There are three general categories of water treatment units: 1) units not intended to prevent, destroy, repel, or mitigate any microorganisms or other pests (e.g., carbon or some other coarse filtering material); 2) units that consist of only a physical or mechanical means of preventing, destroying, repelling, or mitigating any microorganisms or pests (e.g., devices); and 3) units that incorporate a chemical antimicrobial agent or units that consist of a combination of physical and chemical treatment intended to prevent, destroy, or mitigate microorganisms or pests (e.g., pesticides).

Products in the first category are subject to neither registration nor regulation under the Federal Insecticide, Fungicide, and Rodenticide Act (FIFRA). Products in the second category are only subject to regulation under FIFRA. Products in the third category are subject to both registration requirements and regulation under FIFRA.

There are approximately 147 registered water treatment products. The first registration was issued in 1965 by the U.S. Department of Agriculture. The remaining products were registered by the U.S. Environmental Protection Agency beginning in 1975.

The majority of the water treatment products are registered for use in conjunction with municipally treated or microbiologically potable water. Five of these products are registered for use on untreated or raw water (i.e., water of unknown quality or source). Types of registered water treatment products are as follows:

- Water filters 118
- Filtering media 13
- Replacement cartridges 11
- Water purifiers 5

Prior to 1979, data requirements for bacteriostatic water filters consisted of bacteriological and chemical data. These data requirements were published in the Federal Register (1) as the Interim Requirements for Registration of Bacteriostatic Water Treatment Units for Home Use. Since the promulgation of the conditional registration regulations in 1979, we have only required chemical data demonstrating that no more than 50 µg/l (50 ppb) silver are released into the effluent water. This is because with the promulgation of the conditional registration regulations, microbiological data are no longer required for pesticide products with non-public health related uses. Bacteriostatic water filters are in this category because they can only be recommended for use in conjunction with municipally treated water or water that is already microbiologically potable. The only pesticidal claim allowed for this type of product is that it "inhibits (slows down) the growth of bacteria with the filter medium." Other acceptable claims for bacteriostatic water filters are of an aesthetic nature, such as, "removes chlorine, makes the water taste better, clarifies the water, and filters out suspended particles."

On the other hand, water purifiers fall in the category of pesticide products with public health-related uses because they are recommended for use on raw/untreated water or water of unknown source or quality. Therefore, bacteriological and chemical data are still required for those products. The products currently registered as water purifiers are only for emergency use. They are not registered for use on a continuous basis.

In 1984, EPA formed a task force for the specific purpose of developing definitive guidance and specific test parameters for demonstrating effectiveness of water treatment units claiming to microbiologically purify water under conditions that simulated actual use. This task force, which consisted of 17 people, was chaired by Dr. Stephen Shaub of the U.S. Army Medical Bioengineering R&D Laboratory in Frederick, Maryland. The culmination of the efforts of this task force resulted in the Guide Standard and Protocol dated April 1986 and revised in April 1987.

Our current requirements for microbiological water purifiers consist of data showing effectiveness of the product against bacteria, viruses, and protozoan cysts.

In summary, the requirements for registration of bacteriostatic water filters have not changed since 1979. On the other hand, we now have more definitive guidance and specific testing parameters for products claiming effectiveness as microbiological water purifiers - the Guide Standard and Protocol for Testing Microbiological Water Purifiers.

REFERENCE

1. Federal Register, Volume 41, No. 152, August 5, 1976.

CONTROL OF POINT-OF-USE WATER TREATMENT DEVICES IN CANADA: LEGAL AND PRACTICAL CONSIDERATIONS

Richard S. Tobin
Environmental Health Directorate
Health and Welfare Canada
Ottawa, Ontario K1A OL2
Canada

Introduction

With a natural resource of about 25 percent of the world's fresh water, Canada can be considered a water-rich country. Over 7.6 percent of Canada's surface is covered by water, although it is not always located where water demand is highest. With this tremendous supply of water, about 2,500 communities are served by a water distribution system covering 87 percent of the population (1). Most of the remaining 13 percent of the population are served by private wells or other small private sources. It has been estimated that there are as many as 1.38 million private wells in Canada, serving about 4 million people.

Unfortunately, this seeming abundance has resulted in reckless use of water. While Canadians consume only 1.3 L of water per person per day (0.3 gpd) (2) the average rural household uses about 150 L/person/day (40 gpd), and the average demand for municipal systems averages about 500 L/person/day (132 gpd).

Water Jurisdictions

In Canada, the legislative base is derived from the Constitution Act of 1981 (including previous Constitution Acts, referred to in this act). Although the acts do not specifically address water, the ownership of natural resources, including water, is vested with the provinces. The provinces, therefore, have the right to enact legislation with regard to water and to have exclusive jurisdiction over municipal institutions, local works and undertakings, and other matters within the province. Under the Department of National Health and Welfare Act, this department has a responsibility to investigate and conduct programs related to public health. In conducting this program, under Section 5 of the act, the department must coordinate its efforts with those of the provinces. For example, although there is no national safe drinking water act, the Guidelines for Canadian Drinking Water Quality are developed by a Federal-Provincial Subcommittee that reports to a Federal-Provincial Advisory Committee on Environmental and Occupational Health (in turn reporting to the Conference of Deputy Ministers of Health). Thus, the provinces assume the lead role in ensuring a safe supply of drinking water whereas the Federal government provides leadership in ensuring guidelines for drinking water quality.

In some circumstances, the Federal government is entirely responsible for the provision and quality of drinking water. These include administering potable water regulations for all common carriers (transportation crossing Canadian Interprovincial and International borders), and on Canadian coastal shipping vessels, and the provision of potable water in the Territories, Indian reservations, national parks, and military bases.

Under the Food and Drugs Act, administered by Health and Welfare Canada, a "food" is defined (in Section 2) as including "any article manufactured, sold or represented for use as a food or drink for man, ..., and any ingredient that may be mixed with food for any purpose whatever." It is an offense under this act to sell a food that contains harmful or poisonous substances, that is unfit for human consumption, or that has been prepared under unsanitary conditions. The Minister, therefore, has the authority to prescribe regulations for water, although this has only been done for bottled and spring waters (Division 12, Food and Drug Regulations), due to the primary role generally assumed by the provinces.

Point-of-Use Devices

Point-of-use devices are becoming common household appliances in Canada, as they are in the U.S. Total sales of all types of devices have been estimated at about 100,000 per year. By means of a telephone survey of over 16,000 homes in cities across Canada, we learned that the use of activated carbon filters alone ranged from about 0.3 percent to 16 percent (3), largely depending upon the perception of the water quality. Interestingly, the highest value was found in a city with a notorious taste and odor problem, but with no particular chemical problem.

In Canada, there is no specific legislation controlling the sale, use, or performance of point-of-use water treatment devices. A number of acts, however,.do have provisions that could be related to these devices. Some of these are briefly described below.

The Pest Control Products Act (1968-69, c. 50, S.1), administered by Agriculture Canada, regulates products that are intended to control any kind of pest. A pest is "any injurious, noxious, or troublesome insect, fungus, bacterial organism, virus, weed, rodent or other plant or animal pest, and includes any injurious, noxious or troublesome organic function of a plant or animal" (Section 2 of the act).

An exemption (Regulations, Section 3(a)) to the act stipulates that it may not be applied when the product is a food. Since "drink" is defined as a food in the Food and Drugs Act, an administrative agreement between Agriculture Canada and Health and Welfare Canada has exempted water treatment chemicals and devices from consideration under the provisions of the act (4). The agreement states that for "chemicals or devices for water purification," the Environmental Health Directorate of Health and Welfare Canada should be consulted. In fact, water treatment chemicals are not currently covered by legislation at the Federal level, except for emergency disinfection chemicals which may be considered as food additives or drugs (depending on claims made) under Regulations of the Food and Drugs Act. Thus, it is conceivable that such devices (e.g., ozonators, chemical feeders) may be covered by the Act when they are used for nonpotable water (e.g., swimming pools, spas) but not for potable water.

The Medical Devices Regulations (established by PC 1976-2031 and revised periodically) of the Food and Drugs Act are used to control devices that are "manufactured, sold or represented for use in ... prevention of disease ...." Thus, depending on the exact claims made, devices may come under the provisions of these regulations. If the device claims to disinfect water and prevent enteric disease, for example, it could well be interpreted to fall under these provisions. The regulations require the manufacturer to notify the department when a device is put on the market and to furnish certain information including a statement of purpose of the device, and a copy of instructions. The department may also require evidence of the safety and effectiveness of the device (Regulations, Section 27(1)). Thus far, these provisions have not been used for point-of-use devices.

The Hazardous Products Act (1968-69, c. 42, S.1), administered by Consumer and Corporate Affairs Canada (and naming Health and Welfare Canada in certain sections) authorizes the Minister to carry out investigations and demand information regarding consumer products, to determine whether such products are likely to be a danger to the health or safety of the public. Where it is considered necessary to remove a product from the market entirely, it is included in Part I of the Schedule to the Act. Where it is considered that specific regulations can be prescribed to which the products must comply in order not to present a hazard, then the products are listed in Part II of the Schedule.

In 1981, it was proposed to prohibit the sale of activated carbon water filters because of the problem of bacterial growth on the filters (5). As a result of negotiations with industry, an agreement was reached whereby such filters would be labeled to prevent their use on microbiologically unsafe waters. Subsequently, the proposal to ban these devices was discontinued (6).

Another piece of legislation administered by Consumer and Corporate Affairs Canada is the Competition Act (R.S., c. C-23, S.1; 1986, C-26, S.19) which supersedes the Combined Investigation Act. Parts of this act have been used where misrepresentation of devices has been alleged.

Section 36(1) states that

> "No person shall ... (a) make a representation to the public that is false or misleading in a material respect; (b) make a representation to the public in the form of a statement, warranty or guarantee of the performance, efficacy or length of life of a product that is not based on an adequate and proper test thereof, the proof of which lies upon the person making the representation; ... the general impression conveyed by a representation as well as the literal meaning thereof shall be taken into account in determining whether or not the representation is false or misleading in any material respect."

Anyone found guilty of such an offense on conviction or indictment is subject to a fine in the discretion of the court or to imprisonment for five years, or both.

Since these provisions require the representations made for a device to be true and backed by adequate and proper proof, it is clear that these are powerful tools against misleading advertising for all types of devices. The Department has worked closely with officials of Consumer and Corporate Affairs Canada and lawyers in the Justice Department by advising them on technical matters and suggesting test protocols for testing of devices. In many cases we have been asked to be prepared to serve as expert witnesses in the event that the case went to court, which in most cases was not required. Often a guilty plea was entered by the defendants, obviating the need for a trial.

Nonlegislative Activities

Although these legislative tools are at our disposal for more serious problems, the departmental policy has been to avert problems before they occur. Thus, our program on point-of-use devices includes the following aspects: testing and evaluation of devices; provision of advice and educational materials; and cooperation with industry and nonprofit organizations.

In our program of testing devices, summarized elsewhere (7), we have elucidated a basic test protocol and have applied it on a number of water-disinfectant devices, including ultraviolet (UV), filtration, and iodine-releasing devices. Bacterial growth studies have been conducted on several types of activated carbon water filters: normal granular activated carbon, silver containing, compressed carbon, precoated carbon, etc. Results from these studies were qualitatively similar; all types of carbon filters supported bacterial growth, and these bacteria contaminated the finished water.

Evaluation of devices is conducted on an ongoing basis at the request of the public, governmental and nongovernmental agencies, and industry. Normally an evaluation involves the review of data and claims made for a device and provision of an opinion on the suitability of the device for a given purpose, the validity of claims, adequacy of the test protocols, etc. Often we find that there is not sufficient good evidence on which an evaluation can be based.

Advice on the choice and use of devices is often sought by members of the public. Sometimes individual advice is required, but often the educational Tearsheets, Dispatches, Information Letters, Environmental Health Directorate Reports, and scientific articles can be sent to the person for in-depth study.

A number of cooperative efforts have been made with industry associations and nonprofit organizations. For example, numerous discussions have been held with the Canadian Water Quality Association to discuss perceived problems with the advertising and promotion of water treatment devices. They have published voluntary guidelines for use by the industry for the advertising and promotion of carbon water filters (8) and all products (9). In another area, we have worked closely with the National Sanitation Foundation (NSF), who has assumed a lead role in development of performance standards and the testing and listing of devices. Departmental representatives have served on NSF Working Groups during preparation of the draft UV standard, on the revision of the standards on health effects and easthetic effects devices, and on the Joint Committee on water treatment units and on the Council of Public Health Consultants. It is considered that the NSF listing of devices provides the consumer with an easily identifiable proof of performance for removal of specific contaminants. Ultimately, compliance with these voluntary performance standards should make the selection of an appropriate device more straightforward for the consumer.

Conclusion

Although there is no specific Canadian legislation respecting point-of-use water treatment devices, there are a few pieces of legislation that have been or could be used for particular problems. At the present time less formal methods are generally used to provide information on these devices and to ensure their safety and efficacy in use.

References

1. Anon. National inventory of municipal waterworks and wastewater systems in Canada 1981. Supply and Services Canada, Ottawa, 1981.

2. Environmental Health Directorate. Tapwater consumption in Canada. 82-EHD-80, Health and Welfare Canada, Ottawa, 1981.

3. Tobin, R.S., Junkins, E.A. and Eaton, F.E. Survey of the use of activated carbon water filters in Canadian homes. Can J. Public Health. 76:384-387, 1984.

4. Health and Welfare Canada. Antimicrobial products subject to the Pest Control Products Act and Food and Drug Act. Information Letter No. 536. Health Protection Branch, 1978.

5. Health and Welfare Canada. Point-of-use water treatment devices. Information Letter No. 601. Health Protection Branch, 1981.

6. Health and Welfare Canada. Activated carbon water treatment devices. Information Letter No. 635. Health Protection Branch, 1982.

7. Tobin, R.S. Testing and evaluating point-of-use treatment devices in Canada. JAWWA (In press), 1987.

8. Canadian Water Quality Association. Canadian water filter industry voluntary guidelines for carbon water filter advertising and promotional claims. CWQA, Waterloo, Canada, 1982.

9. Canadian Water Quality Association. Voluntary water quality industry product promotion guidelines. CWQA, Waterloo, Canada, 1984.

THE REGULATION OF WATER TREATMENT DEVICES IN CALIFORNIA

Robert F. Burns
Sacramento, CA 95814

During 1986, the California legislature introduced two legislative measures regarding point-of-use (POU) and point-of-entry (POE) water treatment devices. One measure, Senate Bill SB 2119(1) by Senator Torres, addressed the performance of POU/POE water treatment devices for which a claim relative to the health or safety of drinking water is made. The other measure, SB 2361(2) by Senator McCorquodale, addressed advertising claims made in the sale of POU/POE water treatment devices. Both bills were passed by the legislature, signed by the Governor, and became effective on January 1, 1987.

In general, the two bills were considered to be tough pieces of legislation. A frequently asked question is why the legislature decided to regulate the water treatment device industry. Since the California legislature does not maintain a record of their committee proceedings, one can only speculate why these bills were passed. What is known is that a number of events, preceding the introduction of the bills, had come to the attention of the legislature.

Since the late 1970s, Californians have realized that their ground water sources of drinking water were potentially vulnerable to chemical contamination. In 1978, a large number of wells in the San Joaquin Valley were found to be contaminated with the agricultural fumigant, dibromochloropropane (DBCP). Many of these wells were found to exceed the state's action level for DBCP of 1 µg/l (1 ppb).

During 1980, wells in the heavily populated San Fernando Valley and San Gabriel Valley in Los Angeles County, were found to be contaminated by industrial solvents such as trichloroethane (TCE) and perchloroethylene (POE). These same industrial solvents were also detected in wells in the Santa Clara Valley, which is often referred to as Silicon Valley.

In 1985, the California Department of Health Services (DHS) sampled over 3,000 wells used by large public water systems (over 200 connections) for organic chemical contaminants (3). A significant number (18.3 percent) of the wells sampled had measurable concentrations of one or more organic chemicals. One hundred and sixty five of these wells (5.6 percent) had concentrations of chemicals that exceed the State Maximum Contaminant Level (MCL) or a State Action Level (4). When contamination levels were found to exceed an MCL or State Action Level, public notification was initiated by means of a public news release.

During 1985 and 1986, the newspapers and television stations, particularly in Southern California, frequently reported on drinking water contamination problems. The public was alarmed by these news stories and became very concerned about the quality of their drinking water.

As a result of the increased public concerns about drinking water quality, the bottled water industry recorded a significant increase in sales in 1985 and 1986. The water treatment device industry appears to have experienced a similar increase in sales during the same period. Unfortunately, there were a number of cases of consumer abuse and fraud as a result of the overly aggressive marketing efforts by a few companies selling water treatment devices.

The introduction of SB 2361 by Senator McCorquodale has not been tied to any specific consumer problems. However, the Senator represents the Santa Clara Valley and his office had been contacted by constituents about the marketing techniques used by the water treatment industry in that area.

SB 2361 enacted a statute which provides "truth-in-advertising" as it relates to water treatment devices. The statute addresses false or misleading advertising with key provisions which make it unlawful to:

- Make false claims or statements about the quality of water provided by a public water system.
- Make false claims about the health benefits provided by the use of a POU/POE water treatment device.
- Make any product performance claims unless such claims are based on actual, existing factual data.

- Make any other attempts to mislead the consumer or misrepresent the product.

This statute is expected to deter unscrupulous salespersons and reduce the number of complaints relative to fraudulent sales. The statute will also assist the consumers who are victims of fraudulent sales by providing them with a means to file a criminal misdemeanor action and recover damages. The statute does not assign enforcement responsibility to any specific agency. It is expected that local district attorneys and the State Attorney General will take legal action against companies acting in violation of this new law.

The introduction of SB 2119 by Senator Torres is often described as a response to a consumer abuse problem in McFarland, California. A water treatment device company was reported to have advertised and convinced customers in McFarland that their water treatment device could remove all cancer-causing chemicals. This marketing effort occurred at a time when this community was very concerned about a cluster of childhood cancer cases that were being investigated by state and local health agencies. The company was successfully prosecuted by the State Attorney General and the settlement allowed the consumers to rescind their sales contracts.

SB 2119 enacted a statute which requires that any water treatment device for which a health benefit claim is made, cannot be sold in California unless the device had performance testing that has been certified by the Department of Health Services (DHS). This law further requires DHS to adopt regulations setting forth the criteria and procedures for certification of water treatment devices. These regulations must include appropriate testing protocols and procedures to determine the performance of these devices. The cost of this new program is to be paid for through fees imposed on the applicants. The law also assigned responsibility for enforcement to DHS or local health departments.

This statute outlines a very specific Plan for the regulation of POU/POE water treatment devices (WTDs). The general provisions of this statute include the following:

- The DHS is required to adopt regulations which set forth the criteria and procedures for the certification of WTDs that are claimed to affect the health and safety of drinking water.

- A "water treatment device" (WTD) is defined to mean any point-of-use or point-of-entry instrument or contrivance sold or offered for rental or lease for residential, commercial, or institutional use, without being connected to the plumbing of a water supply intended for human consumption in order to improve the water supply by any means, including, but not limited to, filtration, distillation, adsorption, ion exchange, reverse osmosis, or other treatment.

- No WTD which makes product performance claims or product benefit claims that the device affects health or the safety of drinking water, shall be sold or otherwise distributed unless the device has been certified.

- WTDs which are not offered for sale or distribution based on claims of improvement in the healthfulness of drinking water need not be certified.

- A WTD initially installed prior to the operative date of the statute is not required to be certified.

- The requirement that a WTD be certified does not become operative until one year after the effective date of the regulations.

- The DHS or any testing organization designated by the DHS may agree to evaluate test data in a WTD offered by the manufacturer, in lieu of the requirements of the statute, if the DHS or the testing organization determines that the testing procedures and standards used to develop the data are adequate to meet the requirements of the statute.

- The DHS may accept a WTD certification issued by an agency of another state, by an independent testing organization, or by the Federal government in lieu of its own if the DHS determines that certification program meets the requirements of the statute.

The provisions that are to be included in the DHS regulations were defined in the statute with considerable detail. The provisions that are required or allowed as part of the regulations are as follows:

- The regulations shall include appropriate testing protocols and procedures to determine the performances of WTDs in reducing specific contaminants from public or private water supplies.

- The regulations may adopt, by reference, the testing procedures and standards of one or more independent testing organizations if the DHS determines that they are adequate to meet the requirements of the statute.

- The regulations may specify any testing organization that the DHS has designated to conduct the testing of WTDs.

- The regulations are required to include minimum standards for (a) performance requirements, (b) types of tests to be performed, (c) types of allowable material, and (d) design and construction.

- The regulations are required to include requirements relative to product instructions and information, including product operation, maintenance, replacement, and the estimated cost of these items.

- The regulations may include any additional requirements, not inconsistent with the statute, as may be necessary to carry out the intent of the statute.

Finally, the statute specifies procedures for the enforcement of the act. Key enforcement provisions include the following:

- The DHS, or any local health officer with the concurrence of the DHS, is responsible for the enforcement of the act.

- The DHS may suspend, revoke, or deny a certificate upon its determination that either (a) the WTD does not perform in accordance with the claims for which certification is based, or (b) the manufacturer, or any employee or agent thereof, has violated the statute.

- The act provides that any person, corporation, firm, partnership, joint stock company, or any other association that violates any provision of the act, is liable for a civil penalty not to exceed $5,000 for each violation.

The DHS Public Water Supply Branch (PWSB) has been given the responsibility for the implementation of SB 2119 and is currently developing policy and regulations for the implementation of the certification program. The PWSB has established an informal advisory committee consisting of representatives for industry, water utilities, and consumers to assist in identifying and addressing issues. The following are some of the elements of the program and regulations that are being considered:

- The DHS plans to adopt existing protocols and standards such as those established by the National Sanitation Foundation (NSF).

- A "health or safety claim" will be defined in terms of the Primary Drinking Water Standards adopted by the DHS or the U.S. Environmental Protection Agency.

- Certification of a WTD will be based on specific contaminants for which the manufacture has made a health or safety claim.

- The DHS will not establish a state laboratory to conduct the testing required for certification. The DHS plans to contract with outside laboratories or testing organizations for the testing and other administrative tasks relative to certification.

- The DHS may choose not to accept any manufacturer's data relative to the performance testing that will be required for state certification.

The water treatment device industry is very concerned as to how the WTD certification program will impact the marketing and sales of their product in California. The advisory committee has been very helpful in bringing the industries concerns to the attention of the PWSB. Some of the concerns that have been identified are as follows:

- The failure to accept manufacturer's data would impose a substantial cost on the industry.

- If retesting by a third-party laboratory or testing organization is required, the manufacturers will have to pass on the added expense to the consumers.

- The cost of testing under NSF or equivalent standards will be very expensive.

- The one year grace period in which all testing must be completed may exceed the capacity of State contract laboratory or laboratories designated to conduct WTD performance testing.

- In order to reduce the costs associated with performance testing, consideration must be given to testing approaches such as the use of surrogates and the extrapolation of data whenever possible.

It is evident that the California legislature has given the DHS a difficult assignment. However, the Department is committed to the establishment of a WTD certification program that will serve the needs of the California consumers and still be responsive to some of the unique problems of the water treatment device industry. The DHS is also confident that the California program will not be in conflict with any efforts to establish a national certification program.

REFERENCES

1. SB 2119 (Chapter 1247, Statutes of 1986).

2. SB 2361 (Chapter 1278, Statutes of 1986).

3. Organic Chemical Contamination Of Large Water Systems In California. California Department of Health Services, April 1986, page ii.

4. Drinking Water Action Levels set by the Department of Health Services.

WISCONSIN REGULATION OF POINT-OF-USE AND POINT-OF-ENTRY WATER TREATMENT DEVICES

Loretta Trapp
Department of Industry, Labor and Human Relations
State of Wisconsin
Madison, WI 53707

Wisconsin's involvement in regulating point-of-use and point-of-entry water treatment devices involves five different state agencies. Two of those state agencies, the Department of Justice (DOJ) and the Department of Agriculture, Trade and Consumer Protection (DATCP), have consumer protection sections. Water treatment device manufacturers and dealers only become involved with DOJ and DATCP if their advertising literature or sales practices appear to be false or misleading.

A third state agency, the Department of Health and Social Services (DH&SS), is responsible for recommending enforcement standards for ground water contaminants of public health concern. The enforcement standard may be the actual maximum contaminant level (MCL) set by the United States Environmental Protection Agency or may be below the MCL if scientific evidence for the lower number is presently available but was not considered when the MCL was established. When an MCL has not been set for a contaminant of public health concern, the enforcement standard will establish the upper limit concentration for the contaminant in ground water.

Water treatment device manufacturers and dealers rarely become involved with any DH&SS activities. However, the Department of Natural Resources (DNR), the fourth state agency, uses the enforcement standards to establish whether or not a water supply is contaminated. After a public hearing process, the DNR usually adopts the recommended enforcement standard into its regulations. If a water supply contains a contaminant of public health concern in excess of an enforcement standard, the water supply is deemed contaminated. The DNR develops regulations for methods to be pursued in obtaining pure or noncontaminated drinking water for human consumption.

If a water supply contains a contaminant in excess of an enforcement standard, the DNR requires the owner of that water supply to first seek a naturally safe water supply which can involve:

- Extending a well casing;
- Drilling a new well; or
- Connecting to a public water supply or other noncontaminated well.

The DNR requires all water for human consumption to be noncontaminated. [Department of Industry, Labor and Human Relations (DILHR) regulations essentially require all water going to plumbing fixtures to be noncontaminated.] Point-of-use or point-of-entry water treatment devices used to reduce the concentration of contaminants below the enforcement standard may only be installed upon approval of the DNR. The DNR also has the authority to require sampling and maintenance for these water treatment devices. Point-of-use devices are usually not designed to produce the volume or flow rate of noncontaminated water needed and so at this time are not allowed for use on contaminated water supplies. The DNR also considers point-of-entry water treatment devices at this time to be the last resort or at best an interim solution until a naturally safe water supply can be obtained.

Water treatment device manufacturers and dealers may become involved with DNR regulations if they want their devices to be used to reduce the concentration of a contaminant below the enforcement standard.

The fifth state agency is the Department of Industry, Labor and Human Relations (DILHR), which reviews all point-of-use and point-of-entry water treatment devices for the following:

- Rendering inactive or removing aesthetic and health related contaminants;
- Suitability of construction materials for use with potable water;
- Ability of the device to withstand the pressures to which it will be subjected; and
- Proper installation instructions.

The information that DILHR requires for review is contained in Appendix A.

Presently DILHR has adopted into regulation only one nationally recognized standard, the Water Quality Association Standard S-100. DILHR has proposed additional regulations, a copy of which is contained in Appendix B.

In conclusion, water treatment device manufacturers and dealers will most often become involved with DILHR regulations, and to a lesser extent DNR regulations. Appendix C contains a list of contact people for each of the five state agencies.

Appendix A

APPLICATION FOR PLUMBING PRODUCT REVIEW - REQUIRED INFORMATION FOR WATER TREATMENT DEVICES

A letter, requesting approval, must be submitted by the device manufacturer or distributor. Each letter may contain only one product review request. The following information shall be submitted with each request for product review:

1. Product trade name and model number.

2. Manufacturer's name, address and telephone number.

3. Product engineer's name, address and telephone number.

4. Two copies of sales brochure, catalog and other promotional literature.

5. Written detailed description of the composition and function of device.

6. Detailed assembly drawings.

7. Information regarding marking of device:

   a. Method of marking.

   b. List of marking information on device.

   c. Location of markings on device.

8. Complete installation instructions, including detailed installation drawings indicating all connections between the device and plumbing system.

9. A list and copy of all national standards to which the device, or the device's construction materials, conforms.

10. A list of material specifications if other than construction materials in referenced national standards. Documentation shall also be provided indicating that the construction material is accepted for use with potable water, by the National Sanitation Foundation (NSF) or other national agency.

11. The trade name, scientific name and chemical formula of any chemical used in the device that may be added or leached into the water. A toxicity rating and the source of the toxicity rating must also be provided. Documentation shall be included showing that these chemicals are accepted for use with potable water, by the U.S. Environmental Protection Agency (EPA), U.S. Food and Drug Administration (FDA), National Sanitation Foundation (NSF) or other national agency.

12. A signed report, by an approved testing laboratory or the manufacturer, which concludes that the device functions and performs in accordance with assertions submitted to the department. This report must include but is not limited to the following:

    a. A detailed explanation of the test method(s).

    b. The influent temperature, pH, hardness, total dissolved solids and concentration of contaminants.

    c. The effluent temperature, pH, hardness, total dissolved solids and concentration of contaminants.

    d. The minimum detection concentration of the contaminants that may be achieved by the test method.

    e. Test results of at least one duplicate sample.

    f. Test results of a reagent blank.

    g. Test results of a spiked sample.

    h. The percentage of influent disposed as waste.

    i. An estimate of the error in the test results.

    j. A sample calculation.

    k. Test results proving conformance to referenced national standards

    l. Test results indicating the burst pressure.

    m. Test results indicating the working water pressure and temperature range.

n. The name, address and telephone number of the laboratory.

o. The name of the individual(s) performing the test(s).

13. Disposal requirements of any wastewater, backwash fluid, filter, membranes or other replaceable device components.

14. A graph indicating the pressure loss, in psig, through the device, over the entire flow rate range, in U.S. gallons per minute.

15. An operation and maintenance manual or instructions, including but not limited to the following:

a. Maintenance cycle under given influent conditions.

b. Maintenance procedures.

c. Operating pH range.

d. Operating pressure range.

e. Operating temperature range.

f. Operating flow rate range.

g. Operating total dissolved solids range.

h. Any influent conditions that will adversely affect the stated performance of this device.

Appendix B

PROPOSED DRAFT REGULATIONS

Section ILHR 82.11 (181) "Water treatment device" means a device which:

> Renders inactive or removes microbiological, particulate, inorganic, organic or radioactive contaminants from water which passes through the device or the water supply system downstream of the device.

Section ILHR 84.20 (6) (o) *Water treatment devices.*

1. Water softeners shall conform to WQA S-100.

2a. Except as provided in subpar. b., water treatment devices shall function and perform in accordance with the assertions submitted to the department under s. ILHR 84.10, relating to rendering inactive or removing contaminants.

2b. A water treatment device which injects a water treatment compound into a water supply system shall maintain the compound concentration in the system over the working flow rate range and pressure range of the device.

3. Except as specified in subd. 4., water treatment compounds introduced into the water supply system by a water treatment device shall be listed as an acceptable drinking water additive by a listing agency approved by the department. Listing agencies approved by the department shall include:

a. United States Environmental Protection Agency;

b. United States Food and Drug Administration; and

c. National Sanitation Foundation.

4. A water supply system shall be protected from backflow when unlisted water treatment compounds, which may affect the potability of the water, are introduced into the system. The department shall determine the method of backflow protection. Water supply outlets for human use or consumption may not be installed downstream of the introduction of an unlisted water treatment compound.

6. Water treatment devices designed for contaminated water supplies shall be labeled to identify the following information:

a. The name of the manufacturer of the device;

b. The device's trade name; and

c. The device's model number.

Appendix C

PEOPLE TO CONTACT

Department of Agriculture, Trade & Consumer Protection
Trade & Consumer Protection Division
Consumer Protection Bureau
801 W. Badger Road
Madison, WI 53713
Jane Jansen, Director
(608) 266-8512

Department of Health & Social Services
Health Division
Community Health & Prevention Bureau
Section of Environmental and Chronic Disease Epidemiology
1 West Wilson, Rm. 318
P.O. Box 309
Madison, WI 53701-0309

Henry Anderson, M.D., Section Chief
(608) 266-1253

Department of Industry, Labor and Human Relations
Safety and Buildings Division
Office of Division Codes and Applications
201 E. Washington Avenue
P.O. Box 7969
Madison, WI 53707
Loretta Trapp, Plumbing Product Review.
(608) 266-2990

Department of Justice
Legal Services Division
Consumer Protection Bureau
123 W. Washington Avenue
Madison, WI 53703
Kevin O'Conner, Assistant Attorney General
(608) 266-2426

Department of Natural Resources
Environmental Standards Division
Water Supply Bureau
Private Water Supply Section
101 S. Webster
P.O. Box 7921
Madison, WI 53707
William Rock, Section Chief
(608) 267-7649

HOUSEHOLD WATER QUALITY EDUCATION: THE COOPERATIVE EXTENSION SYSTEM ROLE

G. Morgan Powell
Cooperative Extension Service
Kansas State University
Manhattan, KS 66506

Water quality is one of the national priority initiatives for the Cooperative Extension Service System. This paper addresses household water quality education through the Kansas Cooperative Extension Service. It focuses on education relating to one portion of a broader statewide water quality subject. To help you understand why Kansas household water quality programs may not be transferred directly to other states, I will begin with a discussion of the extension system organization using some specific examples from Kansas.

THE COOPERATIVE EXTENSION ORGANIZATION

The extension system in this country consists of the 50 state Cooperative Extension Services and Extension Service of the U.S. Department of Agriculture (USDA). The 1862 land-grant universities, the 1890 land-grant universities, and Tuskegee Institute operate the state Extension Services. These total 67 State Cooperative Extension Service organizations. These extensions are generally funded jointly by Federal, state, and local sources, which explains the name Cooperative Extension Service (two states use the term "Agricultural Extension Service").

Federal funds come through the Extension Service USDA, while state funds come through the respective university systems. Typically, Federal and state funds support the state and area offices consisting of directors, administrators, subject matter specialists, technicians, and support staff. State and local funds support county offices, which consist of county agents, paraprofessionals, and support staff. Local funds come from local governments, usually counties, but sometimes including cities/counties.

The Cooperative Extension Service is the informal, noncredit education arm of the land grant universities. Much of this education occurs at the county level. Subject matter specialists at state and area levels provide support to county extension agents, paraprofessionals, and volunteer teachers and leaders. Because education programs cover broad areas including agriculture, home economics, 4-H, horticulture, and community development, the support must also be broad based. The Cooperative Extension Service includes a very broad range of disciplines to support county education programs. Table 1 shows the number of full time specialists by program area and discipline for Kansas.

Table 1. Extension Subject Matter Specialists in Kansas

	Discipline Specialists	Program Area Specialists
Agriculture		
Agriculture Economics	20	
Agriculture Engineering	8	
Agronomy	13	
Animal Science	16	
Entomology	10	
Grain Science	3	
Horticulture	5	
Plant Pathology	5	
Veterinary Medicine	2	
Sub Total		82
Forestry	12	12
Home Economics	20	20
Community Development	9	9
Energy	6	6
4-H and Youth	13	13
Information (writers, editors, radio, TV, etc.)	21	21
Sub Total		81
Total		163

COUNTY EXTENSION PROGRAM

The county extension education effort helps people:

- Understand, evaluate, and solve problems;
- Learn through informal, out-of-school education opportunities; and
- Work together to develop personally and develop leadership skills.

To achieve these goals, county programs include a wide variety of efforts to reach their clientele. These efforts include public meetings, workshops, short courses, symposia, tours, and demonstrations. Extension programs also reach the public through radio, television, newspapers, and newsletters.

Extension also works closely with clubs and organizations such as 4-H clubs, Extension Homemaker Units, conservation tillage clubs, agricultural commodity clubs, and marketing clubs in Kansas. However, extension also has a history of working closely with the Farm Bureau, soil conservation districts, drainage districts, land improvement contractors, rural water districts, and many other similar groups.

In Kansas, more than 300 county agents conduct county extension programs (Table 2). Kansas also has thousands of volunteer teachers and leaders who add immeasurably to the extension program. A recent survey showed nearly 38,000 part-time volunteers serving in extension sponsored and related organizations, equal to 1.5 percent of the population.

Table 2. County Professional Staff in Kansas

Agricultural Agents	113
4-H/Youth Agents	37
Home Economic Agents	127
Horticulture Agents	12
Total Agents	289
Paraprofessionals	21
Total County Staff	310

KANSAS WATER QUALITY SITUATION

Kansas has relatively few serious water quality problems. However, it does have conditions that are cause for concern and that need careful monitoring. Central water systems roughly serve 80 percent of Kansas residents. Over half of this supply is from ground water. For two years, the Kansas Department of Health and Environment has checked public supply wells for volatile organics and pesticides that are mandated by new regulations of the Safe Drinking Water Act to be implemented from 1987 to 1991. The state has shut down 52 (2.9 percent) of 1,800 wells checked because of contamination.

The Kansas Department of Health and Environment, in cooperation with Kansas State University, randomly surveyed 104 private farmstead wells. They found nitrates in 28 percent of the wells, selenium in nine percent, and fluoride in two percent, where the inorganic contaminants exceeded the MCL (Table 3). They found pesticides and volatile organic chemicals (VOCs) respectively in eight and two percent of the wells (Table 4). These organics are worrisome because they are man-made and the result of fairly recent activity. This suggests that the presence and concentration of organics in private as well as public water supply wells may be increasing and will likely be of more concern in the future.

Table 3. Inorganic Contaminants in Farmstead Wells - Parameters Abvove Maximum Contaminant Level (MCL)

	Percent	Confidence*
Nitrate	28	
Selenium	9	
Fluoride	2	
Lead	2**	
Total Inorganic		37 + 9

* 95 percent confidence level.
** Lead was found to be a result of plumbing in the well water.

Table 4. Organic Contaminants in Farmstead Wells

	Percent	Confidence*
Atrazine	4	
2,4-D	1	
2,4,5-T	1	
Tordon	1	
Chlordane	1	
Heptachlor Epoxide	1	
Alachor	1	
Wells with Pesticides**	8	±6
1,2-Dichloroethane	1	
Benzene	1	
Wells with VOC	2	±3

* 95 percent confidence level.
** Two pesticides were found in each of two wells.

According to the 1980 census, Kansas has about 125,000 private water supplies, almost all wells. Based on 3.9 persons per well, the numbers served by private wells in the survey, roughly 500,000 people (about 20 percent of the state's population) depend on private water supplies. No regulations or testing requirements apply to these private water systems. Users or owners are responsible for the quality of these supplies. They are the operators and the sanitarians. An optimistic estimate is that owners test only a few of these wells each year.

Another concern in Kansas is the large number of abandoned wells that remained unplugged. The Kansas Department of Health and Environment estimated that the state has 250,000 abandoned wells. We believe Kansas could have 500,000 or more abandoned wells.

At Kansas State University, we have determined that a coordinated extension program is essential to address the problems related to private water supplies and agriculture's impact on water quality, especially that of ground water. The Kansas Department of Health and Environment is also concerned about, and supports, this extension education effort.

WATER QUALITY TASK FORCE

The Kansas Cooperative Extension Service initiated a five-member Water Quality Task Force in 1985. This task force began as a component of the agricultural programs portion of extension. The goals were to address the impact of agriculture on water quality and to initiate publications and other educational material that would be needed. In 1986, the task force was expanded to involve other extension program areas. Ten persons representing agriculture, home economics, 4-H, and youth and community development now serve on this task force.

The task force is an efficient and effective way of bringing persons from different disciplines and responsibilities together for discussion, information sharing, program planning, and task assignments. Some persons as they became involved on the task force were reluctant participants. However, as they learned more about specific water quality problems, effects on people, and how corrective action can be taken, these people became enthusiastic participants. Task force members are all involved in preparing extension publications, education programs, and water quality training.

HOUSEHOLD WATER QUALITY PROGRAM

The Extension Household Water Quality Program at Kansas State is coordinated through the Water Quality Task Force. It involves preparation of a wide range of extension publications and supporting media (video and slide/tape) materials, agent and lay leader training, and news stories for newspaper, magazine, radio, and TV.

To date, 12 extension publications are completed and 30 others planned. These publications plus materials from several other sources now make up the Household Water Quality Resource Notebook.

We conducted 10 agent training classes for more than 150 people in October and November 1987. We trained agriculture, home economics, and 4-H/youth agents as well as health services personnel in household water quality.

These one-day training sessions familiarized people with private household water quality and acquainted them with the resource notebook, agency contacts, and how to use them. Our goal was to give local professionals a background so they can be local resources for those with water quality problems.

SUMMARY

The extension system is a complex organization of the 50 state Cooperative Extension Services and the Extension Service, USDA. Federal, state, and local sources fund extension programs. The extension system is the noncredit informal education arm of the land-grant universities. It includes substantial professional staff but many volunteer teachers help to make extension a dynamic and important educational tool for adults and youth.

Kansas has water quality problems among the 125,000 private water wells that serve 20 percent of its population. A recent survey found that 37 percent of the wells exceeded the MCL for inorganics, and 10 percent contained organic contaminants. Although, no data were collected on bacteriological contaminants, based on data from some counties, they could also be substantial. We expect that at least half of the state's private water supplies would not meet standards established by the Safe Drinking Water Act and amendments.

The Kansas Cooperative Extension Service conducts an educational program on household water quality. It provides information on quality of water from private water wells and in all homes. It includes training of county agricultural, home economics, 4-H and youth agents, and county health services personnel as local sources of information. New extension bulletins and leaflets address water quality, water testing, water treatment, and water quality protection. Other resource materials include video tape and slide/tape sets and news programs (radio, TV, newspaper, and newsletter).

Safe drinking water is an important issue. Until more people have safe water, this will likely be an increasingly important issue. Our household water quality program addresses the questions, helps people evaluate their problems, and shows them how to seek solutions.

FEDERAL TRADE COMMISSION REGULATION OF WATER TREATMENT DEVICES

Joel Winston
Federal Trade Commission
Washington, DC 20580

The Federal Trade Commission (FTC) is a small independent Federal agency established in 1914. It has a broad mandate: to promote free and open competition and protect consumers from unfair and deceptive practices. The FTC's mission includes the regulation of anticompetitive practices such as price fixing and monopolization, as well as regulation of business practices that deceive consumers, including false advertising. The FTC works with private and governmental consumer agencies including the Better Business Bureaus, U.S. Postal Inspectors, state attorneys general, and other Federal agencies.

Generally, the FTC can act only in the broad public interest, i.e., when a problem affects large groups of consumers or causes significant harm to smaller groups. Advertising regulation by the FTC tends to focus on national or regional ad campaigns.

The FTC has strong remedial powers when it finds a practice to be unfair or deceptive. While it cannot impose criminal penalties, it can issue broad cease and desist orders to prevent recurrence of the violations or similar practices. In some cases the FTC can obtain Federal court orders enjoining violations, imposing civil penalties, and/or requiring the wrongdoer to make restitution to deceived consumers.

In determining whether an advertisement is deceptive, the FTC first looks at the ad itself to determine what message it conveys to consumers. That message may be explicit or implicit. The FTC then determines whether the ad is likely to deceive reasonable consumers. Both affirmative misrepresentations and omissions of fact may be deceptive. In either case, the FTC will act only when the representation or omission is material, i.e., likely to affect consumers' purchase decisions.

The FTC also enforces its advertising substantiation doctrine. Advertisers making objective claims about their products must have a reasonable basis for the claims prior to making them. The type and amount of substantiation that is required will depend on several factors, including the type of claim, the product, the consequences if the claim is false, the benefits if the claim is true, the cost of developing substantiation for the claim, and the amount of substantiation experts in the field believe is reasonable.

The FTC does not have the resources to pursue every potentially deceptive practice. The FTC considers the following factors in deciding whether to exercise its discretion to act in a particular case:

- The seriousness of the deception;
- The extent of consumer injury, including physical and economic injury;
- The number of consumers misled; and
- The type of product involved. i.e., whether it is a low-cost, repeat-purchase or high-cost, infrequently purchased item.

The FTC has no specific regulations governing the advertising of water treatment devices. Like other advertising, claims for these devices must be truthful and substantiated. Claims for these devices are often credence claims. This means that consumers are not able to evaluate the truth of the claims themselves. For example, a representation that a water treatment device protects users from the hazards of chemical pollutants cannot be evaluated by consumers. Under these circumstances the FTC scrutinizes the advertising more closely.

In recent months, the FTC has brought two formal actions against advertisers of water treatment devices. In the summer of 1987, New Medical Techniques, Inc. was charged with false advertising of its Aquaspring Home Water Distiller. The company made a variety of claims about the capabilities of the Aquaspring -- that it produces pure water, that it will remove all contaminants from the water, and that it will remove all chemicals. In fact, the FTC alleged, the devices were not capable of filtering volatile organic chemicals (such as chloroform and benzene), many of which may be hazardous to health. The company agreed to sign a consent order which prohibits the deceptive claims and requires certain affirmative disclosures in future advertising.

In August 1987, the FTC issued an administrative complaint against North American Philips Corporation, maker of the Norelco Clean Water Machine, a table-top activated carbon filter device. The company's advertisements allegedly represented that the Clean Water Machine would make tap water clean or cleaner, and would help remove organic chemicals. According to the FTC complaint, the machine actually introduced a potentially hazardous organic chemical, methylene chloride, into the water it treated. The company is charged with false advertising and failing to disclose the alleged methylene chloride contamination. The case is scheduled for trial before an administrative judge in early 1988.

In addition to these cases, the FTC continues to monitor advertising for water treatment issues. They are assisted in this effort by the EPA, with which they share authority in this area. EPA has, and continues to provide, scientific expertise to enable the FTC to evaluate and prosecute cases.

In general, the types of advertising claims that the FTC may be concerned about in this area are as follows:

- Claims that a device will purify the water or remove all contaminants. These claims should be supported by reliable evidence and appropriately qualified.

- Claims that ordinary tap water is hazardous to health. These claims should be carefully substantiated and qualified.

- False claims of consumer or expert endorsements. Consumer testimonials should be representative or qualified clearly.

POU/POE PRODUCT PROMOTION GUIDELINES AND CODE OF ETHICS

Maribeth M. Robb
Water Quality Association
Lisle, IL 60532

INTRODUCTION

To gain perspective on our subject, we have to go back about 50 years, when the point-of-use/point-of-entry (POU/POE) water quality improvement industry was born. It was hardly the time to be developing a new consumer product. The country was still in the depths of the depression, and economic prospects in any industry were uncertain. Founding an emerging industry in water quality, in a country known for some of the best water in the world, made the venture even more speculative.

But, those early entrepreneurs saw a niche they thought needed filling: to take what many considered perfect water and tailor it to the specific, individual needs of the user. So they took their ideas and their products into homes and businesses. They demonstrated product effectiveness and, through aggressive marketing, the POU/POE water quality improvement industry evolved.

The success of the industry in the decades since has validated the judgment of those pioneers. For them and for POU/POE, pounding the pavement, talking to customers, and marketing their products were dominant activities for many years.

As the industry grew, so did the number of members representing it. The need for training and education also grew as the many purveyors of POU/POE needed to keep pace with the rising sophistication of the industry. Industry-sponsored organizations, like the Water Quality Association (WQA), were born to help fill that need. WQA continues to fulfill that function today.

ANSWERING CONSUMER CONCERNS

The influences of the marketplace began to change. Because new testing techniques permitted scientists to determine very low levels of toxicity, the quality of the nation's water supply was being reassessed on a daily basis in our newspapers. Many Americans became concerned about the quality of their drinking water. So more products were added to the original lines to meet those consumer needs as well. The industry took on yet another dimension.

Today, experts disagree on the seriousness of the water quality problem. Some argue that the chemical compounds are in such minute quantities that they pose little or no risk to health. Others are more concerned. They worry about long-term exposure to many of the chemicals now detectable.

Even though the issue is unresolved, the fact remains that many individuals find their water unacceptable: too hard, cloudy, smelly, or funny tasting. They may have health concerns about water quality for their family, their infants and young children, and during pregnancies or times of special illnesses.

As a result, there is a growing demand for home treatment systems and for bottled water. Due to this demand, the water quality improvement industry has been thrust from its traditional role of aesthetic water treatment into the role of reducing health-related contaminants. This, in turn, has raised new questions about promotional claims made for various industry products.

VOLUNTARY INDUSTRY PRODUCT PROMOTION GUIDELINES

The promulgation of the Voluntary Industry Product Promotion Guidelines and the creation of the Water Quality Industry Review Panel was prompted by concerns expressed to WQA by various agencies of the Federal governments of the United States and Canada, state and provincial enforcement agencies, and members of the water quality improvement industry. They questioned the general level of industry advertising and promotional claims, and expressed the view that the ads often fall below acceptable norms of accuracy and completeness.

Although WQA was not necessarily in full agreement with these opinions, it nevertheless believed it should respond to them on behalf of the industry in a positive and effective manner. It was hoped that this response would also stimulate companies in the industry to undertake a thorough and comprehensive review of their promotional material. There is plenty of evidence that this occurred and that this activity is ongoing.

The guidelines are not intended to provide all or any part of the wording of anyone's specific promotional material. They are merely designed to provide a general framework within which more accurate and informative advertising, promotional, and sales presentation material can be prepared in such a way as to avoid misleading consumers about the capabilities of water quality improvement products.

Companies in the POU/POE industry are fully aware that they cannot, acting either through WQA or otherwise, agree on matters relating to the form or content of their promotional material or their policies in these areas.

The industry recognizes that promotional material, including advertising and sales presentation material, is a key element in competition and must be left to each individual company to develop for itself.

EPA SUPPORT

From the beginning, the U.S. EPA encouraged WQA to develop a private sector program to address this issue. In a letter dated and signed by three EPA officials, we received much needed support.

> "Over the past ten years, we have experienced a variety of questionable advertising and sales claims by manufacturers or salespeople of water treatment units.
>
> "Unfortunately, the 'wildest' claims have come to be associated with the water quality industry in the mind of public water supply professionals.
>
> "Yet, this is probably not a true characterization of the major companies operating in the water quality field. It would appear that WQA's program is a substantial forward step in correcting this false image.
>
> "Perhaps one of the greatest contributions of the WQA Review Panel would be the preparation of short 'state-of-nation' assessments of advertising and sales claims at the start of the project and periodically in such a way as to enhance public understanding and knowledge. It will also be worthwhile to have an industry forum to which questionable advertising claims can be referred.
>
> "We realize that the WQA Panel will be trying to cope with a difficult problem, but its importance emphasizes the need for the work. Please be assured of our support for your efforts."

GUIDELINES PROVISIONS

The guidelines were adopted in March of 1985 and revised in April of 1987.

The painstaking process which led to these guidelines assured the participants of the following:

- Complaints that are based on factual data;
- Product performance and benefit claims that are verifiable;
- Visuals that are clear and unambiguous;
- Prohibition of untrue, misleading, deceptive, fraudulent, or falsely disparaging claims;
- Prohibition of sweeping, absolute statements;
- True and accurate advertisements;
- Inclusion of pertinent facts;
- Avoiding confusing terminology;
- Performance claims that are based on fact; and
- Problem/solution scenarios that enumerate circumstances and specifics.

Other provisions are spelled out in the handling of:

- Warranties, guarantees, equivalent terms;
- Layouts and illustrations;
- Asterisks;
- Abbreviations;
- Comparisons/disparagement of competition; and
- Testimonials and endorsements.

GUIDELINES REQUESTS

Two types of requests can be submitted through the Voluntary Industry Product Promotion Guidelines process. They are a Complaint Request, filed on an existing advertisement or promotion which can be voluntarily resolved or can move on the Review Panel; and an Advisory Request, in which a company submits its promotional material or advertising for compliance prior to printing or releasing it.

As you can see from Figure 1, communication among involved parties is important to this process from the very beginning. In fact, many of the Complaint Requests submitted are voluntarily resolved before they leave the Staff Review Committee.

REVIEW PANEL

For those that do progress to the Review Panel, the complaint is heard by a highly qualified and conscientious independent panel. This panel forms the "teeth" of the program. Their experience, impartiality, and thorough consideration of each request give the program credibility and consistency.

The guidelines are specific on the credentials of the Review Panel members. It calls for:

Figure 1. Procedures for POU/POE voluntary product promotion guidelines.

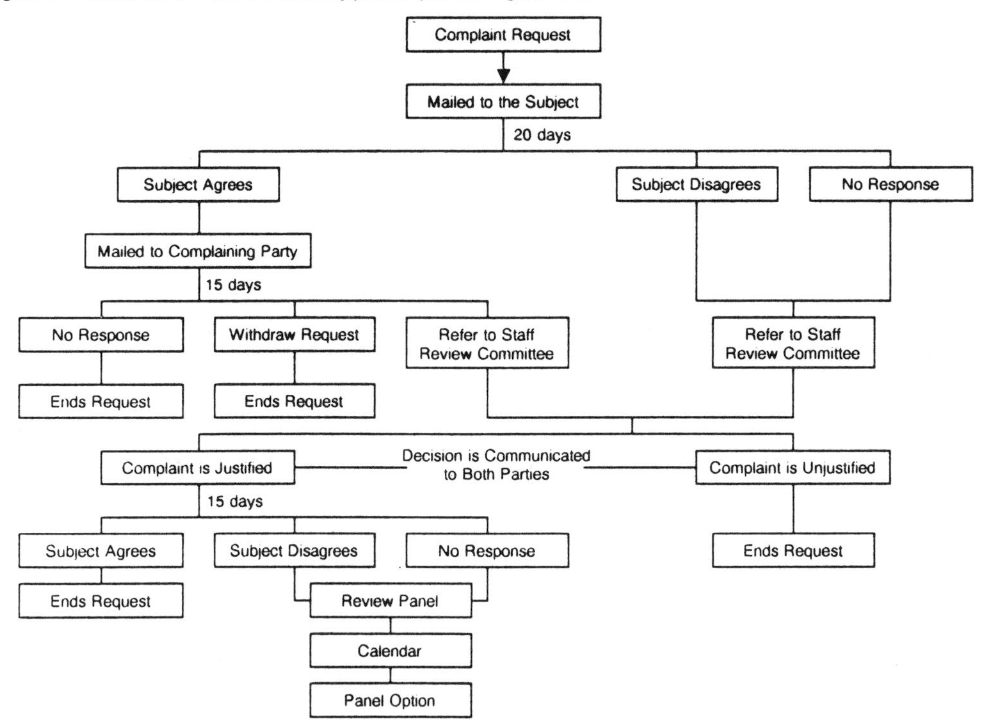

- *A citizen/consumer* - Currently that post is filled by Linda Elwell, who brings her background and experience in direct consumer marketing to her position;

- *An educational institution-connected water chemist* - This position is well served by Roger E. Machmeier, Ph.D., P.E., of the University of Minnesota;

- *A water treatment equipment specialist--nonindustry* - Nina McClelland, Ph.D., brings years of experience with regulation, validation and testing to bear on her panel considerations;

- *A water treatment equipment specialist--industry, but no longer actively employed by industry member* - Wes McGowan now works as a consultant and brings years of industry experience to his panel decisions; and

- *A current former Federal or state government regulatory person who has or is working on consumer and/or misleading advertising problems* - Ron Graham of the Better Business Bureau of Minnesota, Inc. combines his technical prowess with a working knowledge of small-business operations.

COMPLAINT REQUEST CRITERIA

The Complaint Requests heard by the panel are never frivolous. The criteria includes:

- Name, address, and telephone number of source material company;
- Written and dated entries;
- Copy of all materials to be reviewed;
- Details--how, when, where the material was used;
- Particular guidelines possibly violated;
- Explicit request for opinion;
- Full legal name, address, and telephone number of submitting party; and
- The requesting party agrees to not misrepresent and to limit reference to opinion or results.

Figure 2 represents the first two years of the program to date. As you can see, the total number of Complaint Requests increased by two in the second year of the program. However, during the same period, those cases voluntarily resolved increased by four and the total number of Advisory Requests increased by three.

Figure 2. Voluntary guidelines history.

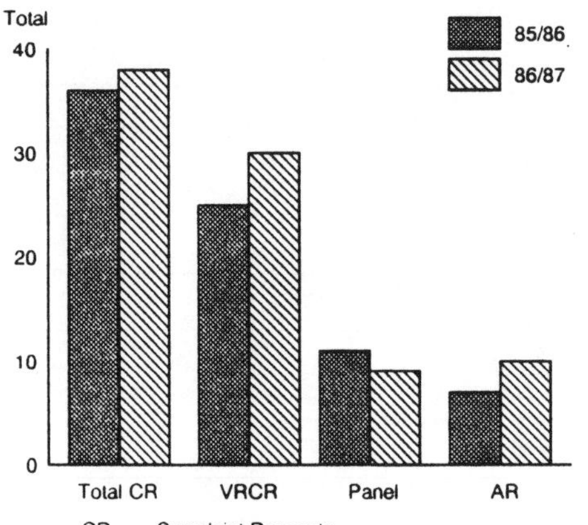

VOLUNTARY COMPLIANCE

What we seem to be seeing is a trend toward voluntary compliance, either after a complaint has been filed or before the material is released. These are encouraging figures for the industry and for consumers as well.

There is also a question of perspective that I would like to stress here. Of the millions of dollars spent by the industry on advertising every year, of the thousands of promotional pieces and the hundreds of advertisements, fewer than 80 have been submitted to the Voluntary Industry Promotion Guidelines Review Program during its two years of operation. A substantial proportion of those advertising materials that did reach the Review Panel were submitted by their own companies to assure compliance prior to publication.

So, changes are being made, and they are being felt in the marketplace.

CODE OF ETHICS

Building on the encouraging response to the Voluntary Industry Product Promotion Guidelines, industry members have now taken another courageous step. Using the guidelines as a nucleus, they have launched a Voluntary Industry Code of Ethics program.

Although it is strictly voluntary, it is hoped that the marketplace, which fueled this industry from the start, will see the benefits of trading with Voluntary Industry Code of Ethics signature companies, and will embrace it as a stipulation for doing business.

The formal hearings procedure for the Code of Ethics will be the same as for the Voluntary Industry Product Promotion Guidelines.

The Voluntary Industry Code of Ethics will be published in early 1988, with the first list of subscribers-in-good-standing published in the second quarter of 1988.

LOOKING AHEAD

Times have changed, and the point-of-use/point-of-entry water quality improvement industry has changed as well.

That change is apparent from marketing, the Voluntary Industry Product Promotion Guidelines and Code of Ethics, and Product Validation.

Water quality is no longer a simple issue. It demands a complex approach to problem solving as well as a blending of industries and specialties to assure optimum quality for all consumers, in all regions of the country.

The point-of-use/point-of-entry water quality improvement industry is poised to meet that challenge with accountability, confidence, and technical expertise.

We have the industry products, the industry people and the drive. By working with professionals like yourselves -- the regulators, the educators, the resource people, and the water utility managers -- we can provide the solutions demanded and needed by consumers today. Together, we can bring economical, customized, quality water to every tap in this country.

NSF's LISTING PROGRAM FOR POU/POE DWTUs

Randy A. Dougherty
National Sanitation Foundation
Ann Arbor, MI 48106

The National Sanitation Foundation (NSF) was chartered in 1944 in the State of Michigan as a private, independent, not-for-profit organization. The mission of NSF is to develop and administer programs relating to public health and the environment in areas of service, research, and education. NSF is best known for its consensus standards and third-party certification programs.
The subject of this article is NSF's listing program for point-of-use and point-of-entry drinking water treatment units (POU/POE DWTUs).

NSF standards are consensus standards and, as such, are part of the public domain. A company can self-certify to NSF standards, or another organization may certify them. The standards specify the minimum requirements for a product to satisfy public health concerns. Additional requirements that a company must meet to obtain and maintain authorization for listing and use of an NSF Listing Mark are specified by contract and administrative policy. The key elements of NSF's listing programs (how NSF certifies initial and continuing conformance to NSF standards) are:

- NSF standards;
- Registered Listing Marks;
- Public listings information;
- Evaluation and testing by an independent, objective third-party;
- Monitoring;
- Corrective action; and
- Enforcement.

NSF STANDARDS

NSF standards are voluntary, consensus standards developed by a Joint Committee (comprised of regulators, users, and industry representatives), reviewed and accepted by the Council of Public Health Consultants (CPHC), and reviewed and formally adopted by NSF's Board of Trustees.

The standards are developed or revised by the Joint Committee (or task groups appointed by the Joint Committee) with the active participation of public health and other regulatory officials, users, and industry.

The role of the CPHC is to assure that the requirements of a standard satisfy public health concerns. CPHC has 36 members from Federal, state, and local regulatory agencies in the United States and other countries, and academia. Industry is not represented. The expertise of the members includes public health, medicine, chemistry, toxicology, epidemiology, microbiology, and engineering. CPHC reviews and must accept a standard or revision before it is sent to the board for adoption; however, it does not make any changes to a standard. If the council does not accept a standard, it is sent back to the Joint Committee. The CPHC is not a "rubber stamp." For example, the council rejected proposed Standard 55 (for ultraviolet drinking water treatment systems) in 1985 because it did not satisfy public health concerns about cysts, turbidity, and viruses.

The Board of Trustees reviews the standards for business and legal consideration. As with the CPHC, the board does not make changes in a standard. If the board does not accept a standard, it is sent back to the Joint Committee.

NSF has the only standards for POU/POE DWTUs that have widespread recognition and acceptance by public health officials. These are:

- Standard 42: Drinking Water Treatment Units - Aesthetic Effects;

- Standard 53: Drinking Water Treatment Units - Health Effects; and

- Standard 58: Reverse Osmosis Drinking Water Treatment Systems.

Shortly after adoption of Standard 53 in 1981, Canadian authorities proposed a ban on point-of-use carbon units because of concern for bioaccumulation - specifically, of opportunistic pathogens. NSF organized and hosted a meeting of

Canadian and U.S. regulatory representatives, manufacturers, and recognized expert consultants. It was reported at the meeting that no known illness could be traced to carbon filters. There was general agreement to add to the standards, and to the labels on carbon units, the following statement: "Activated carbon filter units covered by this standard are not intended to be used where the water is microbiologically unsafe or with water of unknown quality." A revision to Standard 53, to incorporate this statement and additional labeling requirement, was adopted in June 1982. Standard 42 was first adopted in 1973, but was revised in June 1982 to be a companion document to Standard 53. Standard 42 is consistent with Standard 53, but is for aesthetic claims. Although these standards (42 and 53) are not limited to specific treatment technologies, the test protocols are appropriate for carbon or mechanical filtration units only.

Standard 58, for reverse-osmosis systems, was adopted in November 1986.

There are three other standards being developed:

- Proposed Standard 44: Cation Exchange Water Softeners;
- Proposed Standard 55: Ultraviolet Disinfection Systems; and
- Proposed Standard 62 (for distillation systems).

The standards are for units designed to be used for the reduction of specific contaminants from public or private drinking water supplies. The standards have detailed requirements and protocols for testing the units to verify claims for the reduction of specific chemical, particulate, or microbiological contaminants, bacteriostasis, disinfection (Standard 55 only), or for the addition of polyphosphates or silicates (Standard 42 only).

While the primary focus is verification of water treatment claims, the standards include requirements for materials, design, and construction of units to assure that:

- Materials in contact with the drinking water do not impart toxic substances, taste, odor, or color to the water; and
- The units accomplish the intended purpose when installed and operated in accordance with the manufacturer's instructions.

For DWTUs with water treatment claims for the reduction of contaminants that are established or potential health hazards, the standards include requirements for performance indicators, warnings, or other means to alert the user when the unit is not functioning properly. This may be by the DWTU having a shut-off to terminate the discharge of treated water, sounding an alarm, 50 percent reduction in flow, or by providing a test kit. For carbon units, one alternative is to have a 100 percent safety factor, which is verified by testing to twice the rated capacity. For reverse osmosis systems for nitrate reduction, Standard 58 requires either a nitrate monitor on the unit, or the manufacturer must provide a test kit for nitrates with the system.

The standards also have detailed requirements for installation and operating instructions, dataplate information and labeling, and other information about the function and capability of a unit, including specific warnings for users.

NSF LISTING MARK

NSF Listing Marks are formally registered with the U.S. Patent Office, and in Canada. NSF owns the mark, but doesn't use it on products. The mark is for use by other companies, on listed products and in conjunction with listed products, as authorized by NSF. A company applies for and contracts with NSF for authority to use an NSF mark; and specifically agrees to use the mark on only new products fully complying with all NSF requirements. NSF has legal, contractual, and ethical obligations and responsibilities to monitor and verify that only authorized companies use the mark, and use it properly. This is the basis of NSF's authority relating to listed products.

To be considered listed, a DWTU must bear the NSF Listing Mark, and must also bear a model number (or serial number) that distinguishes it from nonlisted units. Consumers and regulators can look for the mark as evidence that a unit is listed by NSF.

The listing program for DWTUs differs from other NSF listing programs in that listed units do not have to meet the same requirements - we verify the specific water treatment claims made by the manufacturer for a unit. So the water treatment claims must be directly associated with the mark. The listing mark with an example of verified water treatment claims is shown in Figure 1.

PUBLIC LISTINGS INFORMATION

One goal of listing services is to make current listings information readily available and easily accessible. This is achieved by publishing and widely distributing seven annual listing books (see Table 1). NSF also publishes up to nine supplements to each annual book. These supplements include complete listing information for new companies, and for companies that cancel listing services; for revised listings, the supplements have the changes only. Because listings change daily, it is impossible to provide current listings information by publication. Therefore,

Figure 1. NSF listing mark on a DWTU.

Listed under NSF Standard 53 for the reduction of TTHMs, Cysts, and Turbidity *only*.

Listed under NSF Standard 42 for the reduction of Taste, Odor, and Chlorine *only*.

Caution: Do not use where the water is microbiologically unsafe or with water of unknown quality, without adequate disinfection before and after the unit.

Table 1. NSF Listing Publications

Title	Distribution
Food Service Equipment and Related Products, Components, and Materials	6,500
Plastics Piping System Components and Related Materials	3,600
Drinking Water Treatment Units and Related Products, Components and Materials	3,200
Swimming Pools, Spas, and Hot Tubs Circulation System Components	3,000
Special Categories of Equipment, Products, and Services	2,800
Wastewater Treatment Units and Related Products and Components	2,400
Class II Biohazard Cabinetry	1,600
Total	23,100

beginning in January 1987, NSF provides for direct electronic access by computer. Any person with a compatible computer and modem can apply for this service and directly access official listings information, which is updated daily. The only cost to a user for this service is the cost of the telephone call.

Another goal is to provide listings information that can be useful to someone selecting a DWTU; therefore, the listings for DWTUs include the following information: the company name and address, a description of the unit or system, the model number of the DWTU and replacement element, and the function (the verified water treatment claims). The listings information also includes the service cycle or capacity in gallons, flow rate, and other information.

EVALUATION AND TESTING BY AN INDEPENDENT, OBJECTIVE THIRD-PARTY

NSF evaluates and tests products as an independent, objective third-party. As a third-party, NSF serves the interests of regulators and users, as well as industry. NSF has five regional offices, four in the continental United States, and one in Brussels, Belgium. NSF regional personnel visit each production location (point of final assembly or production) to evaluate products and select samples for laboratory testing. NSF has its own modern, state-of-the-art laboratory in Ann Arbor, Michigan, and provides a full range of chemical, physical/performance, and microbiological testing. (NSF also has procedures for qualifying and authorizing other laboratories as subcontract or alternate testing laboratories for testing products for listing by NSF.)

MONITORING

Listing is on an annual basis. The listing program is not a one-shot deal, but an ongoing program with continuous monitoring of listed products' conformance to NSF standards. The listing program for DWTUs includes a requirement for annual unannounced inspections by our regional personnel to verify that there have been no unauthorized changes in materials, components, design, or production of listed units. NSF also requires periodic retesting (at least once every five years) of listed DWTUs.

NSF investigates complaints of noncompliance of listed units. The complaints may be from public health or other regulatory officials, or from users. NSF also investigates complaints from other manufacturers - a mechanism for effective self-policing by industry. In all cases, NSF conducts its own investigation, and takes action with the manufacturer if, and only if, NSF confirms that the product doesn't comply.

CORRECTIVE ACTION

The goal of the listing program is to assure that a listed product conforms to an NSF standard. If NSF determines that a listed unit does not conform with the standard, NSF requires the listed company to take appropriate corrective action, which may include the following:

- Evaluation or testing to quality changes to products,
- Modification of equipment,
- Destruction of product,
- Product recall, and
- Public notice.

ENFORCEMENT

NSF zealously strives to maintain the integrity and credibility of the mark; therefore, for repetitive or serious noncompliance, NSF takes specific enforcement action as follows:

- Increased monitoring,
- Administrative hearings,
- Delisting,
- Legal action, and
- Cancellation of contract.

CURRENT STATUS OF THE LISTING PROGRAM AND FUTURE DEVELOPMENTS

The listing program for DWTUs is small. As shown in Table 2, there are only 13 companies with listed units. But, this is one of the fastest growing listing programs. Twenty-one additional companies have applied for listing; and two additional standards (Standard 44 for water softeners and Standard 55 for ultraviolet systems) are expected to be adopted by the end of 1987. Although the program is small, it has the third largest distribution of listing publications (see Table 1), which is an indication of the interest and importance of this program to regulatory officials and consumers.

One exciting item is the development of a "model compound" concept for testing carbon units for volatile organic compounds (VOCs). Under the guidance of a Standard 53 task group, NSF and a listed company have developed isotherm data and dynamic testing data which demonstrate that chloroform can be used as a satisfactory model compound for verifying the reduction of a number of specific VOCs (regulated and nonregulated) by carbon units. This will result in reduced testing costs for verifying contaminant reduction claims for a large number of organic chemicals, producing increased participation by industry and an increased number of listed models. But of even greater importance, as new organic contaminants are found in drinking water supplies, it may be possible to demonstrate that chloroform is a satisfactory model compound for verifying effective reduction by carbon units - which means that there may be a large number of listed units immediately available as a remedy.

Table 2. Listing Service Programs (9/1/87)

Program	Number of Standards	Listed Companies
Food Service Equipment	21	1,033
Plastics Piping System Components	1	275
Swimming Pools, Spas, and Hot Tubs	1	49
Class II Biohazard Cabinetry	1	12
Drinking Water Treatment Units	3	13
Wastewater Treatment Units	4	12
Flexible Membrane Liners	1	3
Special Categories	5	31
Total	37	1,428

SUMMARY

The listing programs are voluntary. But a listing program becomes more than voluntary when NSF standards are referenced in regulations or codes. Regulations or codes usually do not require that a product be listed by NSF, but a listed product is usually accepted by the responsible regulatory agency. By voluntarily participating in the listing program, with required testing, retesting, and unannounced plant inspections by a third-party, a company demonstrates the intent and capability to manufacture a product conforming to an NSF standard. The advantage to a company is wide acceptance of its listed product(s) by regulatory officials and consumers.

Regulatory officials and consumers have assurance that a credible, objective third-party, widely recognized by public health officials, has actually tested and verified that listed products comply with specific standards; and the cost of the program is placed in the private sector rather than adding to the cost of official regulation.

WATER QUALITY ASSOCIATION VOLUNTARY PRODUCT VALIDATION PROGRAM AND VOLUNTARY CERTIFICATION PROGRAM

Lucius Cole
Water Quality Association
Lisle, IL 60532

PRODUCT VALIDATION

One of the basic purposes of the Water Quality Association (WQA) is to promote the acceptance and use of point-of-use/point-of-entry industry equipment, products, and services. One of the most successful programs to promote this concept was the development of voluntary industry standards.

Consistent with the goals expressed in WQA's corporate charter, which are "to foster the further development of equipment, products and services in the industry for the purpose of providing a better way of life for all mankind," the procedures set out in both the Guide for Product Standards Development and the Guide for Product Validation Program were established and are followed in the development of voluntary industry standards. The WOA developed five voluntary industry standards:

- S-100-85 for household commercial and portable exchange water softeners;
- S-101-80 for efficiency-rated water softeners;
- S-200-73 for household and commercial water filters;
- S-300-84 for point-of-use low pressure reverse osmosis drinking water systems; and
- S-400-86 for distillation drinking water systems.

Under this program, a manufacturer may voluntarily submit a specific type of equipment or system to the WQA laboratory where its performance will be carefully evaluated in accordance with the appropriate standard. When a system has successfully performed to the specific testing protocol, it is then qualified to receive the appropriate "gold seal." This seal alerts the consumer that the equipment he/she is considering purchasing complies with the specifications of a very rigorous testing program. At the present time, over 300 products produced by companies in the water quality improvement industry have been validated by the WQA laboratory. A directory of these validated water conditioning products is published semi-annually, and is available to both consumers and regulatory officials.

A 10-page brochure published by the Council of Better Business Bureaus, Inc., entitled *Tips on Water Conditioners*, makes the following statement:

> "When choosing a water conditioner, look for equipment that bears the gold seal of the Water Quality Association. This seal indicates that the Water Quality Association has judged that the equipment complies with the specifications of the industry standards for water softeners (S-100)."

A great deal of effort has been made by the point-of-use/point-of-entry industry to provide the consumer with products that are reliable and perform to basic standard requirements. It has also been the industry's desire to have these systems installed in a safe and economic fashion.

PROFESSIONAL CERTIFICATION

The WQA certification program was established by association members to provide industry-wide standards for evaluating the knowledge of point-of-use/point-of-entry water treatment personnel and to improve the knowledge of those who service the consumer, thereby encouraging professionalism and integrity in the industry. Since the inception of WQA's certification program in 1977, nearly 1,800 people have been certified in one of three categories: dealer, specialist, or installer.

In order to provide support for industry-wide standards for evaluating knowledge, it was vital to develop proper study materials. The water treatment fundamentals correspondence course was prepared to provide comprehensive study materials for the dealer or specialist. This course was divided into 12 lessons. Each lesson consists of a pre-lesson questionnaire to evaluate one's knowledge about the subject, exercises to help evaluate one's knowledge while preparing a lesson, and finally, questionnaires that are completed and mailed to the WQA headquarters for correction. The corrected questionnaires are then returned to the student for his review. This study material has been widely used by individuals who desire to raise their level of technological competency in the point-of-use/point-of-entry industry, wish to participate in the WQA certification program, and want to be able to display the coveted certification emblem.

The necessity and function of various water conditioning systems are well recognized. Equally important to the consumer is the correct installation of the equipment into the plumbing system, since the equipment usually connects to a potable water supply and to an appropriate drainage system.

Improper installations are not only hazardous, but also costly when corrections must be made. In order to prevent such problems from occurring, it is essential that the installer be knowledgeable in acceptable installation procedures as well as local code requirements.

The purpose of the WQA installers home study course is to provide installers with generally acceptable installation procedures relating to point-of-use/point-of entry water conditioning systems. With such knowledge, an installer may perform the installation procedures correctly so that a safe and efficient installation is made. Similar lesson procedures as previously discussed for the fundamentals course are used to assist and evaluate the student's knowledge.

A national directory of certified personnel is published every two years listing the three categories of certification - certified dealers, certified specialists, and certified installers. Each individual listed has completed a specific study course and demonstrated his knowledge by successfully passing the appropriate examination. As a certified individual, he agrees to maintain high standards of service. The Water Quality Association Certification and Education Committee may revoke an individual's right to use the seal of certification if evidence of failure to maintain these standards is established.

The members of the Water Quality Association continue to demonstrate their interest in serving the consumer with both certified specialist/ installer programs and validated products. WQA members over the last 32 years have provided funding in excess of $250,000 for the development and implementation of five WQA standards and six NSF standards to meet the association's goal "to foster the development of equipment, products and services by the industry for the purpose of providing a better way of life for all mankind."

GUIDE STANDARD AND PROTOCOL FOR TESTING MICROBIOLOGICAL WATER PURIFIERS

Stephen A. Schaub
U.S. Army Biomedical Research and Development Laboratory
Frederick, MD 21701

Charles P. Gerba
Department of Microbiology and Immunology
University of Arizona
Tucson, AZ 85721

INTRODUCTION

Over the past several decades, a number of water equipment manufacturers have developed technology for the removal of chemical and microbiological constituents from waters to be used for personal consumption. The need for this capability arises principally from consumer interest in improving the quality of untreated or partially treated waters like those used by hikers, campers, recreational home and boat owners, and families or communities having individual home or small system water sources.

One of the major concerns in water treatment is the need to remove pathogenic microorganisms (bacteria, viruses, protozoa, fungi, and helminths) from the water before its consumption, since it is recognized that infectious disease transmission by water is a significant public health concern.

It is important that water treatment units or devices designed for the protection of human health be effective against pathogenic microorganisms in untreated or partially treated water, and be capable of providing this service over the designed operational life of the equipment in waters likely to be encountered in the United States. These requirements are necessary for protection of the public's health by both the water industry and the government.

A multidisciplinary task force was formed in 1984 to develop a Guide Standard and Protocol for Testing of Microbiological Water Purifiers. The task force was comprised of persons representing the interest areas of academia, industry, and government for research and development, product evaluation and registration, and product regulation and enforcement. The objective of this task force was the development of a standard and protocol that industry, government, and consumers could agree with and which could be attained with current knowledge and technology. The primary emphasis was to protect the consumer.

At this time, the guide standard and protocol has been prepared by the task force, has been technically reviewed (notice in Federal Register of May 29, 1986; and U.S. EPA Science Advisory Panel), and has been appropriately revised in consideration of these reviews. It has been accepted on a provisional basis by the U.S. EPA's Office of Drinking Water and Office of Pesticide Programs, pending experimental verification of the efficacy of the protocol under the prescribed parameters.

The intent of this paper is to provide the major features and considerations of the guide standard and protocol in its current configuration. It is recommended that persons or organizations wishing to use the guide standard and protocol for testing purposes, obtain the complete, detailed package from the above U.S. EPA program offices.

REQUIREMENTS FOR A MICRO-BIOLOGICAL WATER PURIFIER

The current definition of a microbiological water purifier is that it must remove, kill, or inactivate all types of disease-causing microorganisms from water to make the product safe to drink.

Units or devices having limited claims for the treatment or removal of a specific type of organism, or use in a specific, limited application, can be tested for that use in accordance with the protocol, however, such equipment cannot be called a microbiological water purifier. For example, a protozoan cyst removal unit could be tested against the protocol and could demonstrate acceptable cyst removal, but unless it also met the required removals for the bacteria and viruses, it could not claim to be a microbiological water purifier.

PRINCIPLES FOR THE GUIDE STANDARD AND PROTOCOL

The guide standard and protocol is to be considered a general guide, presenting only the minimum features and framework for testing, and may be amended or added to for the evaluation of unique units or specific operational problems (including alternative organisms and procedures) as long as the level of testing and intent of the protocol are not diminished. It is performance-based, utilizing realistic worst case conditions. The goal is to ensure that microbiological requirements of the National Primary Drinking Water Regulations are met by equipment defined as microbiological water purifiers.

The guide standard and protocol is intended to be a living document, subject to revision and update as new knowledge and technology arise. The document should be capable of addressing appropriate test challenges for other types of purifiers when they become available, and would consider new or evolving pathogens of concern if they represent an increased challenge to technology covered by the protocol.

It was intended that the test conditions or requirements of the protocol could be met in reasonably well-equipped laboratories when performed by competent scientists and engineers. It is known that there are at least several commercial and university laboratories in the U.S. that currently have the capability to meet the test requirements.

The protocol does not and cannot address all conceivable microbiological and physical/chemical challenges that could be possible in water. Presently, the test protocol addresses the following technologies:

- Ceramic filtration candles or units (with or without chemical bacteriostatic agents),
- Halogenated resins and units (with or without filtration capabilities), and
- Ultraviolet (UV) units (with or without filtration and/or chemical adsorption capabilities).

MICROBIOLOGICAL CHALLENGES FOR WATER PURIFIERS

The microbiological challenges for testing were chosen to be representative of bacterial, viral, and protozoan pathogens of the gastrointestinal tract, and are believed to cover the treatment requirements presented by most other human pathogens from the gastrointestinal tract or other origins, including fungi and helminths. It is recognized that there are a number of alternative organisms that could have been selected, and which would be equally representative for testing. A detailed rationale for the use and test levels of the challenge organisms is provided in the complete guide standard and protocol, which can be obtained from the U.S. EPA offices mentioned in the Introduction.

Table 1 provides a brief summary of the test organisms and the culture/assay conditions required for testing. In all cases the microbiological procedures chosen represented well documented protocols or standard methods, which could easily be attained in the laboratory.

Table 2 provides the minimum microbiological challenge levels to be used. A major point which must be emphasized is that the challenge levels in most instances exceed the highest concentrations that would be found in typical source waters. It was the task force opinion that the higher challenge levels were less of a concern to the evaluation of purification units than the complications arising from the introduction of analytical errors, which could be introduced from the effluent (product) water sample concentration procedures. Low challenge levels would necessitate sample concentration to quantitatively assay the product waters if microbial removals were significant, especially for the viruses and protozoa.

NONMICROBIOLOGICAL TEST PARAMETERS

It was determined that, in addition to the microbiological challenges to the various water purifiers, there was a need to evaluate the treatment capabilities of units in the presence of associated physical/chemical parameters in water, which may impact on the overall microbial removal capabilities of each type of treatment technology. It was decided that for the first half of the testing procedure a general challenge, typical of most tap waters, would be utilized for all testing. The second half of the testing program would use the worst case challenge in which pH; Total Organic Carbon (TOC); Turbidity; Total Dissolved Solids (TDS); Temperature; and for UV light units, a UV Quenching Test Component, would be added. Table 3 provides the test conditions required for the various types of purifier tests and the recommended materials or chemicals for adjusting the water characteristics. Additionally, silver leaching test conditions for units containing silver bacteriocide are included. While many of the worst case challenges appear to be on the high side of normal conditions, they are not thought to be out of line with conditions brought about by seasonal or meteorological events or significant pollution events in surface waters. The worst case challenges will be maintained over the total duration of the second half of the testing program with the exception of high turbidity conditions, which would be introduced only during sampling periods to prevent excessively rapid clogging of units containing filtration components.

Table 1. Microbiological Methods for Test Protocol

BACTERIA:	
Organism	*Klebsiella terrigena*
Culture Requirements	Overnight broth cultures to obtain stationary growth phase cells
Assay	Spread/pour plate or membrane filter techniques using nutrient agar, M.F.C., or M-Endo Medium*
VIRUSES:	
Organism	Poliovirus Type 1 (LSc) and Rotavirus, Strain SA-11 or WA (both viruses will be tested together using equal proportions to seed the challenge water)
Culture Requirements	Grown on tissue culture and prepared to provide monodisperse virus particles for tests
Assay	Plaque or immunofluorescent foci assays on continuous cell cultures
PROTOZOA:	
Organism	*Giardia lamblia* or *Giardia muris* where disinfection is principal mechanism; 4 to 6 μm spheres can be used where occlusion filtration is the exclusive removal mechanism
Culture Production	Obtain and prepare cysts from feces of laboratory-infected animals
Assay	Count physical particles for filtration: determine viability of cysts (or trophozoites) for disinfectant-containing units

* Use procedures in *Standard Methods for the Examination of Water and Wastewater*, 16th Edition, APHA, or equivalent.

Table 2. Microbiological Challenge (testing according to NSF Standard 53 for cyst reduction will be acceptable)

Organism	Influent Challenge*	Minimum Required Reduction Log	Minimum Required Reduction Percent
BACTERIA			
Klebsiella terrigena (ATCC-33257)	10^7/100 ml	6	99.9999
VIRUS			
Poliovirus 1 (LSc) (ATCC-VR-59), and	1 x 10^7/l	4	99.99**
Rotavirus (WA or SA-11) (ATCC-VR-899 or VR-2018)	1 x 10^7/l		
CYST (PROTOZOAN): *Giardia****			
Girdia muris or *Giardia lambia*, or	10^6/l	3	99.9
As an option for units or components based on occlusion filtration: particles or spheres, 4 to 6 μm	10^7/l	3	99.9

* The influent challenge may constitute greater concentration than would be anticipated in source waters, but these are necessary to properly test, analyze and quantitatively determine the indicated log reductions.
** Virus types are to be mixed in roughly equal 1 x 10^7/l concentrations and a joint 4-log reduction will be acceptable.
*** It should be noted that new data and information with respect to cysts (i.e., *Cryptosporidium* or others) may in the future necessitate a review of the organism of choice and of the challenge and reduction requirements.

Table 3. Test Waters - Nonmicrobiological Parameters

Test Waters	pH*	TOC**, mg/l	Turbidity***, NTU	Temperature, °C	TDS†, mg/l
General	6.5-8.5	0.1-5.0	0.1-5.0	20 ± 5	50-500
Halogen Disinfection Tests					
Chlorine and Others	9.0 ± 0.2	≥ 10	≥ 30	4 ± 1	1,500 ≥ 150
Iodine	5.0 ± 0.2	≥ 10	≥ 30	4 ± 1	1,500 ≥ 150
Ceramic Candle Tests	9.0 ± 0.2	≥ 10	≥ 30	4 ± 1	1,500 ≥ 150
Ultraviolet Tests‡	6.5-8.5	≥ 10	≥ 30	4 ± 1	1,500 ≥ 150
Silver Leaching Tests	5.0 ± 0.2	~ 1.0	0.1-5.0	20 ± 5	25-100

Recommended Materials for Adjusting Water Characteristics:
* Inorganic acid or base.
** Humic acids.
*** AC fine dust (part No. 1543094).
† Sea Salts (Signma Chemical Co. or equivalent).
‡ p-hydroxybenzoic Acid (general purpose reagent).
Quench UV to just above alarm point. (Add color or reduce light intensity to just above point where low UV intensity alarm would be triggered.

PURIFIER TEST PROCEDURES

For testing of a purifier, it is recommended that three units be set up in parallel, according to manufacturer's instructions for normal line pressures (~60 psig), but the flow rates through the units would not be controlled. The required physical/chemical characteristics for the tests would be maintained continuously with the exception of turbidity as mentioned above. The bacterial challenge should be maintained continually during operation of the units, but virus and protozoa would only be introduced during the sampling "on" periods of the test. Samples are to be taken in duplicate from each sampling position (influent and product water) at the appropriate sampling "on" periods specified in the protocol. Sufficient void volumes will be passed through the units before samples are taken to ensure the uniformity of the challenge, especially for the virus, protozoa, and turbidity, which are introduced only during the sampling periods. An exception is sampling after the programmed 48-hour stagnation periods when samples are taken immediately upon start-up. Disinfectant-containing purifier samples (microbiological) will be immediately neutralized with respect to that chemical.

The testing program will be conducted for 100 percent of the estimated treatment capacity of halogenated resin-containing units and for a 10-1/2 day operating period with ceramic candle and UV units. The operating cycles for the purifier units should be representative of use (e.g., intermittent). For example, a use cycle of every 15 to 40 minutes during each operating day with an actual operating "on" period of 10 percent for each cycle is considered appropriate. (For example, if a 30-minute cycle is used, the operating "on" period would be three minutes/cycle.) If necessary, due to time or laboratory constraints, a shorter operational day with an extended test period can be substituted, or a daily operating cycle of 20 percent "on" and 80 percent "off" can be used.

A schematic of a typical test stand for the evaluation of plumbed-in units is presented in Figure 1. This schematic is essentially that of the National Sanitation Foundation (NSF) Standard 53 for Drinking Water Treatment Unit Health Effects. This set-up allows good control of total test operations and sampling with a minimum of variables entering the procedure. Sampling can be performed on an automated basis. Testing of portable or hand-held purifier units can be set up in a batch testing procedure, which would follow the test schematic of plumbed units as closely as possible, although a number of the features of the system such as flow meters, automated sampling procedures, and delivery of virus, cyst, and turbidity challenges would have to be modified.

Tables 4, 5, and 6 illustrate the sampling plans for the various types of units. For units containing halogens, the sampling of residual halogen in the product water is conducted at the same times and frequencies as the microbiological challenges. Additionally, at the start of each test, the waters to be used for testing of all type units are to be examined for U.S. EPA primary and secondary pollutant constitutents in accordance with standard analytical procedures. The challenge conditions for iodine versus chlorine and other halogen-containing units are identical except after the 48-hour stagnation period at 75 percent of the life of the units, wherein the pH challenge for iodine-containing units becomes pH 5.0 rather than 9.0. The sampling for ceramic candles and UV units is straightforward. Leaching tests for silver-containing units are also necessary to make sure that no dangerous levels of silver reach the product water.

MINIMUM MICROBIOLOGICAL REMOVAL FOR ACCEPTANCE OF PURIFIERS

In order to meet the standards of acceptable microbiological removal, the three duplicate units tested must continuously meet or exceed the defined microbiological removal requirements, within allowable tolerances as determined from paired influent and product water samples. Not more than 10 percent of the sample pairs from the three units can fall below the tolerances for removal:

- Bacteria: 99.999 percent removal,
- Virus: 99.9 percent removal, and
- Protozoa: 99.5 percent removal.

If the geometric mean of all sample pairs meets or exceeds the microbiological removal requirements, the deficiencies of the 10 percent of the sample pairs falling below tolerances will be allowed and the purifier capabilities will be considered acceptable.

It is important to keep records of the test procedure and the data if there is a claim to be made for units to be considered microbiological water purifiers.

PRELIMINARY TEST RESULTS OF THE FEASIBILITY OF THE PROTOCOL

Recently, studies have been conducted on cartridge type filters, using the protocol for ceramic candle units to help ascertain the feasibility of the protocol for water purifier testing. The tests were conducted specifically to evaluate the virus testing component of the protocol. Several modifications to the protocol were made to simplify the testing procedure. These included elimination of the in-line mixer and booster pump, elimination of the pack-pressure regulator, addition of a 380-l (100-gal) reservoir to contain the challenge waters, and another similar reservoir to collect the product water for disinfection prior to discharge. The viruses were added to the proper concentration in the reservoir in a batch mode for

Figure 1. **Test apparatus - schematic** (adapted from NSF Standard 53).

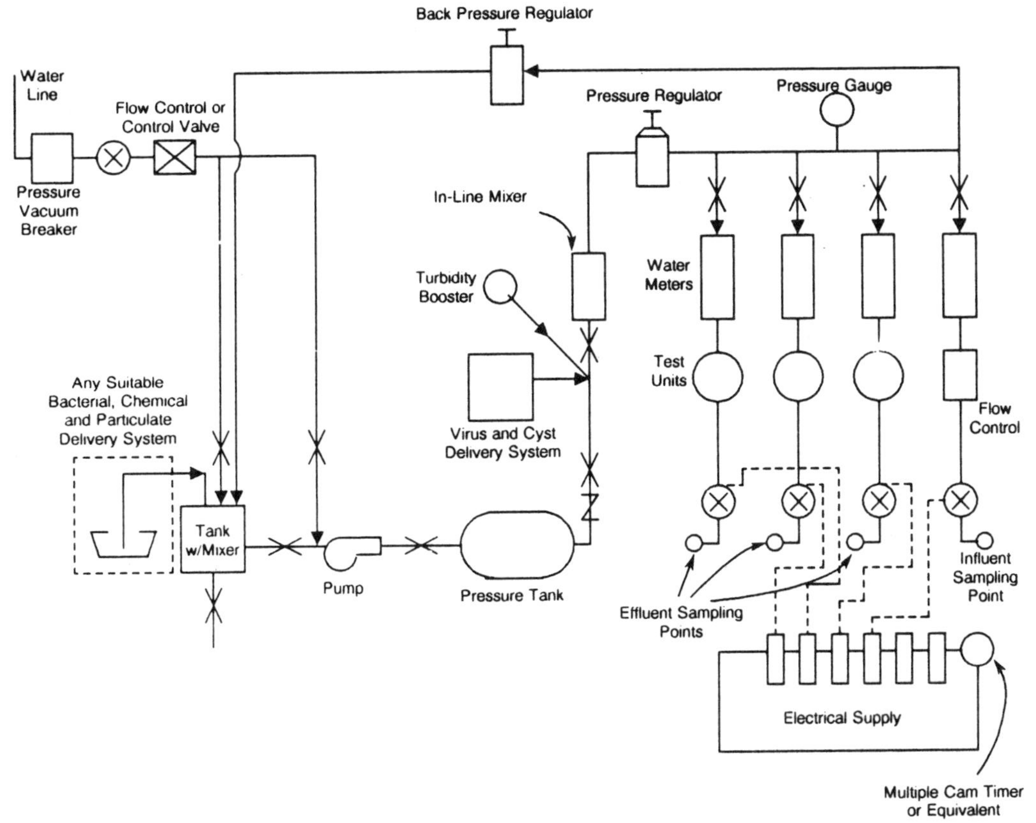

NOTES:

1. Faucets are to be used in testing all units under the sink or over the sink. (Regular kitchen faucets for stationary units and faucet attached units and smaller third faucets for by-pass units.
2. Faucet attached units and portable units are to be placed after the solenoid.
3. Whole house or similar large units need not use faucets. Flow can be regulated with valves placed on the effluent side.
4. All materials of construction must be suitable for use with drinking water.

X Shut Off Valves (not to be used for regulating flow

Z Solenoid Valve

⊗ Check Valve

Table 4. Sampling Plan Halogen-Containing Units

		Tests				
Test Point (as % of total stated capacity)	Test Water	Influent Background (all Halogens)	Residual Halogen, Chlorine & Others	Iodine	Microbiological Chlorine & Others	Iodine
0	General	X	X	X	X	X
25			X	X	X	X
50			X	X	X	X
After 48-hr stagnation			X	X	X	X
60	Challenge		X	X	X	X
75	pH 9.0 ± 0.2		X	X	X	X
After 48-hr stagnation			X	X	X	X
90	Challenge			X		X
100	pH 9.0 ± 0.2 (chlorine & others) or		X	X	X	X
After 48-hr stagnation	pH 5.0 ± 0.2 (iodine)			X		X

Table 5. Sampling Plan - Ceramic Candles or Units and UV Units

		Tests	
Test Point*	Test Water	Influent Background	Micro-biological
Start	General	X	X
Day 3 (middle)			X
Day 6 (middle)			X
After 48-hr stagnation			X
Day 7 (middle)	Challenge		X
Day 8 (near end)			X
After 48-hr stagnation			X
Day 10.5			X

* All days are "running days" and exclude stagnation periods. When the units contain silver, a leaching test shall be conducted as shown in Section 3.5.1.e and silver residual will be measured at each microbiological sampling point.

Table 6. Sampling Plan - Leaching Tests for Silver-Containing Units

	Tests	
Test Point	Influent Background	Silver/ Residual
Start	X	X
Day 2		X
After 48-hr stagnation		X

introduction to the filters. When worst case challenge waters were applied, the virus inoculum was added to the water only at the sampling time. The testing cycle was every 30 minutes for an eight-hour daily run with an operating "on" period of three minutes. Virus samples were taken when approximately 10 bed volumes of the seeded water had passed through the system units except for 48-hour stagnation samples.

The test apparatus is shown in Figure 2. The results of the virus challenge are shown in Table 7. The results indicate that the units can remove at least 99 percent of the virus from the regular challenge (Tucson tap water), which was used in the first half of the test, and greater than 99.9 percent removal from worst case challenge water used in the second half of the tests.

CONCLUSIONS

The Guide Standard and Protocol for Testing Microbiological Water Purifiers provides the water industry, consumers, and government a common approach to the evaluation of existing and developmental products for their microbiological removal capabilities. While the standard and testing protocol is rigorous in terms of both microbiological removal requirements and challenge requirements (both microbiological and physical/chemical), it should provide a high degree of confidence in terms of protection to consumers wishing to use point-of-use microbiological water purification units or devices to remove disease-causing organisms from their drinking water.

Figure 2. Test apparatus.

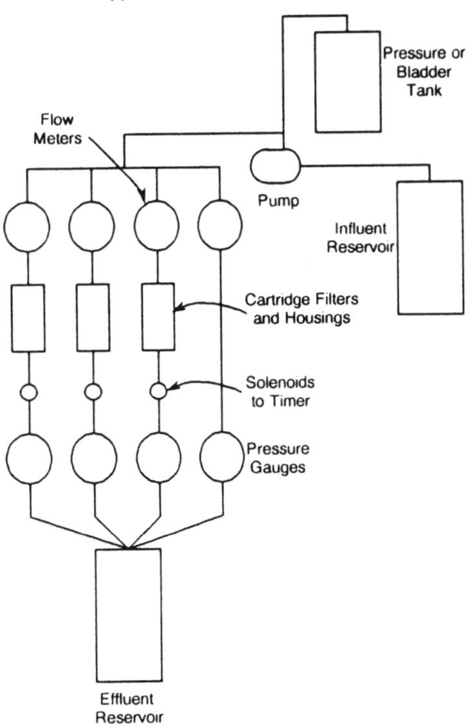

Table 7. Virus Removal by Cartridge Filter Units Using EPA Water Purifier Protocol

Time (days)	Water Type	Average Percent Removed
1	Tap Water	99.32
3	Tap Water	98.33
6	Worst Case Water	99.89
7	Worst Case Water	99.91
8	Worst Case Water	>99.99
10	Worst Case Water	99.93
10	Worst Case Water (stagnant)	>99.99

PERFORMANCE AND APPLICATIONS OF GRANULAR ACTIVATED CARBON POINT-OF-USE SYSTEMS

Karl Van Dyke and Roy W. Kuennen
Research and Development Division
Amway Corporation
Ada, MI 49355

Point-of-use (POU) water treatment devices based on granular activated carbon (GAC) have been around for many years. Most were traditionally used to improve the aesthetic quality of drinking water: color, turbidity, taste, and odor. As the knowledge of trace organic chemicals in drinking water has grown, so has the public's awareness of the problem. As the issue of chemical contamination grows, the traditional aesthetic claims are being supplemented by claims for chemical removal. This requires more clearly defined ways to evaluate the validity of these claims. The addition of the VOCs to the Safe Drinking Water Act, with their low maximum contaminant levels (MCLs), places more stringent requirements on devices claiming to remove them to below MCL concentrations (1,2).

It has been said one can filter anything from water if it is filtered through enough money. This means that the technology is available, however, it may be too expensive to employ on a full scale. Point-of-use water filters offer options of technology that could be prohibitively expensive on a large scale. They offer a means to achieve filtration if properly designed to remove even low µg/l (ppb) levels of contaminants that are below MCLs. POU devices containing GAC are designed in one of three basic configurations: GAC in single or sequential housings, GAC and PAC in sequential housings, and pressed carbon blocks. They may be installed in several configurations including faucet, stationary, faucet diverter, and line bypass.

The purpose of this paper is to summarize results of lab and field studies on POU devices to support the concept of using chloroform as a surrogate compound for making removal claims for specific VOCs found in drinking water.

POINT-OF-USE PERFORMANCE -- TEST DATA

GULF SOUTH RESEARCH INSTITUTE

The first significant evaluation of POU devices began with the Gulf South Research Institute (GSRI) studies sponsored by the U.S. EPA Office of Drinking Water and reported between May 1979 and October 1981 (3). This study consisted of three phases progressing from lab to field. It was set up to develop basic data and information on the performance of a variety of small home treatment units with respect to organics removal and bacterial/endotoxin aspects. The philosophy was to stress the units under simulated home use.

Phase 1

Phase 1 addressed protocol development and pilot testing using spiked and unspiked New Orleans tap water. The basic procedure was to run units on an accelerated program, sampling at several points throughout the rated life. Influent and effluent samples from filter tests were run for trihalomethanes (THMs), nonpurgeable total organic carbon (NPTOC), bacterial standard plate count (SPC), endotoxin, and silver (where appropriate). Unspiked tapwater was selected as the main means of testing based on pilot tests. Preliminary results on seven units in Phase 1 included two faucet filters (one bypass, one nonbypass), one portable pour-through, one stationary filter, and three line bypass POU devices.

Results for trihalomethanes showed:

- Small faucet and pour-through filters removed amounts ranging from negligible to about 25 percent of the influent THM during the manufacturer's recommended filter life.

- Larger stationary and line bypass filters removed greater percentages, ranging from 43 percent to over 90 percent for one filter. The extent of THM removal appeared to be a function of several factors, including the quantity of carbon relative to treated water, contact time, and design factors.

Phase 2

A total of 25 commercially available units and one experimental unit were evaluated in Phase 2. Influent and effluent tests were run for THMs, NPTOC, SPC, endotoxin, silver, and peripheral substances. Each model was challenged with ambient New Orleans municipal drinking water during three replicate tests.

Results for trihalomethanes showed:

- Small faucet and pour-through filters removed from four percent to 69 percent of the influent THM during the rated lifetime of units.
- Larger stationary and line bypass filters removed greater percentages, ranging from 15 to 98 percent. Six line bypass units (including one unit containing an experimental material) removed over 85 percent of the influent THM. The descriptions of the five commercially available line bypass units (Table 1) allows a comparison of rated capacities: from 3,785 to 15,140 l (1,000 to 4,000 gal), carbon weights from 300 to 1,708 grams (0.7 to 3.76 lb), and iodine numbers from 434 to 1,223. Units are ranked based on carbon weight and iodine number, a measure of capacity. Breakthrough curves (Figure 1) show performance as expected with three models performing well for 3,785 l (1,000 gal), and two past 7,570 l (2,000 gal). Removal here should be noted as percent of total chloroform applied to the filter, not influent versus effluent at any one point.

Table 1. Line Bypass Units (Phase 2)

Unit	Description	Rated Capacity (gal)	Carbon Weight (g)	Iodine Number
Culligan SG-2	1 cartridge w/GAC	4,000	1,708	980
Aqualux CB-2	2 cartridges w/GAC	2,000	1,150	966
Everpure QC4-THM	2 cartridges (1 PAC, 1 GAC)	1,000	765	1,223 (GAC) 798 (PAC)
Aquacell Bacteriostatic	2 cartridges w/GAC	2,000	417	867
Seagull IV	1 cartridge w/pressed block	1,600	300*	434*

* Carbon and binder.

Phase 3

Phase 3 included a ground water study, field study, home study, and an addendum covering removal of halogenated organics.

In the ground water study, 10 models were evaluated with well water spiked with 1,1,1-trichloroethane, carbon tetrachloride, trichloroethylene, and tetrachloroethylene at target levels of 50, 20, 50, and 50 µg/l respectively. The five commercial line bypass models previously detailed in Phase 2 were among the 10 selected.

The results were:

- The selected models (from each of the four basic configurations) removed from 76 to 99 percent of the spiked halogenated organics in the well water. The five line bypass models highlighted in Phase 2 removed from 93 to 99 percent of the spiked organics.
- Generally, carbon tetrachloride and 1,1,1-trichloroethane broke through the carbon filter first while trichloroethylene and tetrachloroethylene did not break through at all for some units.

The Phase 3 field study involved units tested in the cities of Miami, Florida; Atlanta, Georgia; Pico Rivers, California; and Detroit, Michigan.

The results were:

- Upon reviewing the THM reduction data, it appears that the relative unit performance ranking determined in the laboratory test is maintained throughout the field study.
- The overall level of specific organic chemical removal appears to be adversely affected by increased levels of background organic material present in the water matrix.

The conclusions drawn were:

- Filter unit percent removals of THM compounds can be predicted with some confidence using results based on laboratory tests. The total organic background level, NPTOC, in water may affect percent reduction levels of THMs slightly (to approximately 10 to 20 percent) over the range of background organic levels experienced in the field study (0.6 to 7.1 mg/l NPTOC).
- The ranking of filter units appears to be maintained regardless of the source water. The ranking could change if one were monitoring the filter's effectiveness in removing a different contaminant, since many carbon adsorbents are manufactured with some degree of selectiveness. The GSRI experience has been that this occurs infrequently and one can be fairly certain that units demonstrating effective removals of THM compounds will also demonstrate strong affinities to most other halogenated compounds.

In the home study, three filter types were challenged with New Orleans tap water under nonaccelerated conditions. One of the line bypass models was included in this test. Limited data indicated the validity of using accelerated laboratory testing to provide an accurate assessment of the effectiveness of carbon filtration in removing trace organic chemical contaminants from drinking water. A larger data base

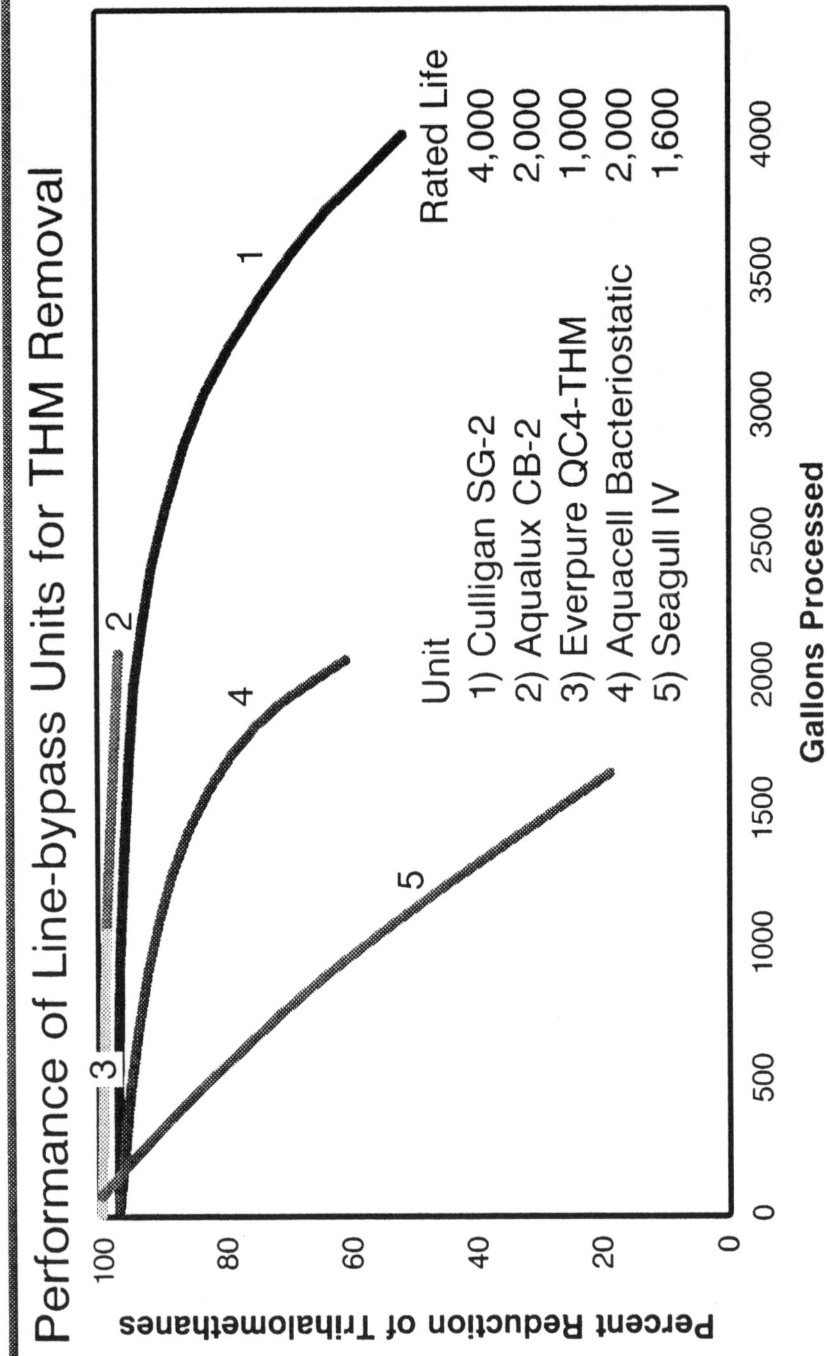

Figure 1. Summary
Performance of line-bypass units for THM removal

would be required to assure the positive correlation between laboratory (accelerated) tests and home (actual use) tests although preliminary data indicate the lab tests provide valid assessment data.

The Phase 3 addendum covering removal of halogenated organics involved testing the same 10 POU devices with additional organic spikes. The study was flawed due to the presence of organic solvents used to get the halogenated organics into water. It was felt that the organic removal data provided useful information regarding the removal capabilities under very severe conditions.

The selected models were effective in removing the spiked level of halogenated organics from New Orleans tap water (49 to 99 average percent removal). Line bypass units, generally, are more effective in removing halogenated organics than the other models.

Line Bypass	65-99 percent
Faucet Mount	50-60 percent
Stationary	49 percent
Pour-through	66 percent

The conclusion was that it was reassuring to observe significant reductions of halogenated organics and THM contaminants (49 to 99 percent and 24 to 99 percent, respectively) despite the adverse conditions of the test water challenge.

EPA/NSF JOINT STUDY

A more in-depth study of POU devices in the home was run using five of the units tested in the GSRI study. An EPA-sponsored study performed by the National Sanitation Foundation (NSF) to study point-of-use reduction of volatile halogenated organics in drinking water involved two communities. The project was conducted in 1983 and 1984, in Silverdale, Pennsylvania, and in Rockaway Township, New Jersey to determine whether point-of-use carbon treatment is cost effective for the control of volatile halogenated organic chemicals in small water systems and also to study water quality district management techniques for point-of-use treatment (4). Criteria for selection of devices for this study included, among others:

- The devices must have demonstrated greater than 95 percent reduction of halogenated organic demonstrated in the GSRI Phase 3 study or equivalent.

- The manufacturers were required to certify that their products met NSF Standard 53 Section 3, for structural integrity, corrosion resistance, nontoxicity, etc.

- Point-of-use devices were required to have a rated capacity exceeding 2,650 l (700 gal) (estimated one-year service life).

Only line bypass models were selected.

SILVERDALE

For the Silverdale study, the summary of influent VOC results (Table 2) covering March 1983 through April 1984 shows trichloroethylene and tetrachloroethylene as primary contaminants, with smaller amounts of carbon tetrachloride and chloroform.

Table 2. Influent VOC Results - Silverdale

Compound	Mean Conc. Predevice (µg/l)
Trichloroethylene	80.4
Tetrachloroethylene	20.6
1,1,1-trichloroethane	1.1
1,2-dichloroethane	< 1.0
Carbon tetrachloride	8.0
Chloroform	6.7
Bromodichloromethane	1.5
Dibromochloromethane	1.4
Bromoform	< 1.0

Breakthrough was defined as detection of the same VOC in consecutive postdevice samples from the same location at a concentration above the routine detection limit of 1.0 µg/l. Breakthrough did not occur for any of the devices tested during the 14 months of sampling for TCE and PCE.

ROCKAWAY TOWNSHIP

For the Rockaway Township study, 12 POU devices were installed on private wells in October 1981. The type and concentration of VOCs was varied, with the primary contaminants being 1,1,1-trichloroethane and trichloroethylene. See Table 3.

Table 3. VOC Results - Rockaway Township

VOC	Range of Conc. Found (µg/l)	Number of Wells w/VOCs
1,1,1-trichloroethane	1.0-240.0	8
1,2-dichloroethane	6.7-20.7	4
Tetrachloroethylene	1.0-12.3	7
1,2-dichloroethane	< 0.4-10.1	6
Trichloroethylene	0.7-240.2	4
Trans-1,2-dichloroethylene	0.8-5.1	2
Chloroform	1.7-2.1	2
Trichlorofluoromethane	< 25.0	1

The local health department did sampling and analysis from October 1981 to October 1982 with 100 percent VOC reduction. After being included in the study, four sites were monitored in October 1983, the

24th month of operation. There were no VOCs measured in postdevice samples after 24 months of operation.

Data from the study indicates that point-of-use granular activated carbon (GAC) treatment devices effectively reduced concentrations of trichloroethylene, tetrachloroethylene, carbon tetrachloride, 1,1,1-trichloroethane, 1,1-dichloroethylene, 1,1-dichloroethane, and chloroform at influent concentrations studied. These results confirm bench and field results from the Gulf South Research Institute study.

NSF STANDARDS 42 AND 53

The National Sanitation Foundation has issued several standards relating to POU devices, the two most relevant being Standard 42 for Drinking Water Treatment Units - Aesthetic Effects, and Standard number 53 - Health Effects (5). The contaminants covered in Standard 42 include such things as taste, odor, and color and specific chemicals such as foaming agents, hydrogen sulfide, and phenol. Particulate reduction, while not an inherent quality of GAC, is also covered.

The contaminants covered in Standard 53 include total trihalomethanes, six pesticides, soluble ions -- nitrate and fluoride, eight heavy metals, plus cysts, turbidity, and asbestos. Total trihalomethanes are run at a 450 µg/l challenge to a 100 µg/l effluent limits. This standard does not yet cover the newly regulated VOCs.

DEFINITION OF END OF LIFE

Determining when a POU device is exhausted for taste and odor can be done by consumer perception. NSF standard determines it by chlorine reduction. For the trihalomethanes (MCL 100 µg/l), the NSF Standard 53 challenges at 450 µg/l and rates capacity based on breakthrough to 100 µg/l, with a 100 percent safety margin. End of rated life for the other VOCs may be based on a breakthrough equal to the MCL with an evaluated challenge of perhaps 300 µg/l. The VOCs and MCLs are listed in Table 4 (2).

Table 4. Regulated VOCs and MCLs

Compound	MCL (mg/l)
Vinyl Chloride	0.002
Benzene	0.005
Carbon Tetrachloride	0.005
1,2-dichloroethane	0.005
Trichloroethylene	0.005
1,1-dichloroethylene	0.007
p-dichlorobenzene	0.075
1,1,1-trichloroethane	0.200

To perform to these levels, a POU device will need to be well designed and constructed.

AMWAY DATA ON A POU WATER TREATMENT SYSTEM

Amway has developed considerable data on POU water treatment system performance. In our laboratories we have been working heavily on evaluation and claims documentation for GAC-based POU devices for the past five years (6,7). The following data is from a unit containing a unique pressed carbon block design. Performance is rated on the conservative position that the only way to prove the filter's ability to remove chemicals from water throughout its rated life, is to test against every chemical claimed. Testing is carried out to 150 percent of rated life to provide an extra margin of safety. The filter is effective for removal of 116 compounds, including 100 of the EPA priority pollutants, plus several pesticides including aldicarb, EDB and DBCP, fuel hydrocarbons, and others.

CHEMICAL CLASSIFICATION

In order to test that many compounds, they were placed into chemically similar classes, then into groups, and then a group was tested together. The classes and groups (Table 5) are based on chemical similarity and analytical technique. Some groups were subdivided for easier analysis.

Table 5. Chemical Classification

Class	Group	Analytical Method
1. Acids	Phenols	EPA-625
2. Base/Neutrals	a) Biphenyldiamines	HPLC-UVD
	b) Chlorinated Hydrocarbons	EPA-625
	c) Cyclohexenone	EPA-625
	d) Haloethers	EPA-625
	e) Nitro compounds	EPA-625
	f) Phthalates	EPA-625
3. Hydrocarbons	Gasoline/Kerosene/Diesel Fuel	GC-FID
4. PBBs	PBBs	HPLC-UVD
5. PCBs	Aroclors	EPA-608
6. Pesticides	a) Halogenated Alkanes	GC/ECD
	b) Nitrogen/Phosphorous	GC/NPD
	c) Organochlorine	EPA-608
	d) Organonitrogen	HPLC-UVD
7. PNAs	Polynuclear Aromatics	HPLC-UVD
8. Purgeables	a) Aromatics	EPA-624
	b) Halogenated Alkanes	EPA-624
	c) Halogenated Alkenes	EPA-601
	d) Trihalomethanes	EPA-601

TEST PROTOCOL

The test protocol is similar to the GSRI and the NSF test Protocol (6). All testing was done on duplicate devices. The main test stand (Figure 2) provides control of the critical parameters. The contaminants are injected under control of an HPLC pump. This

Figure 2. Main test stand.

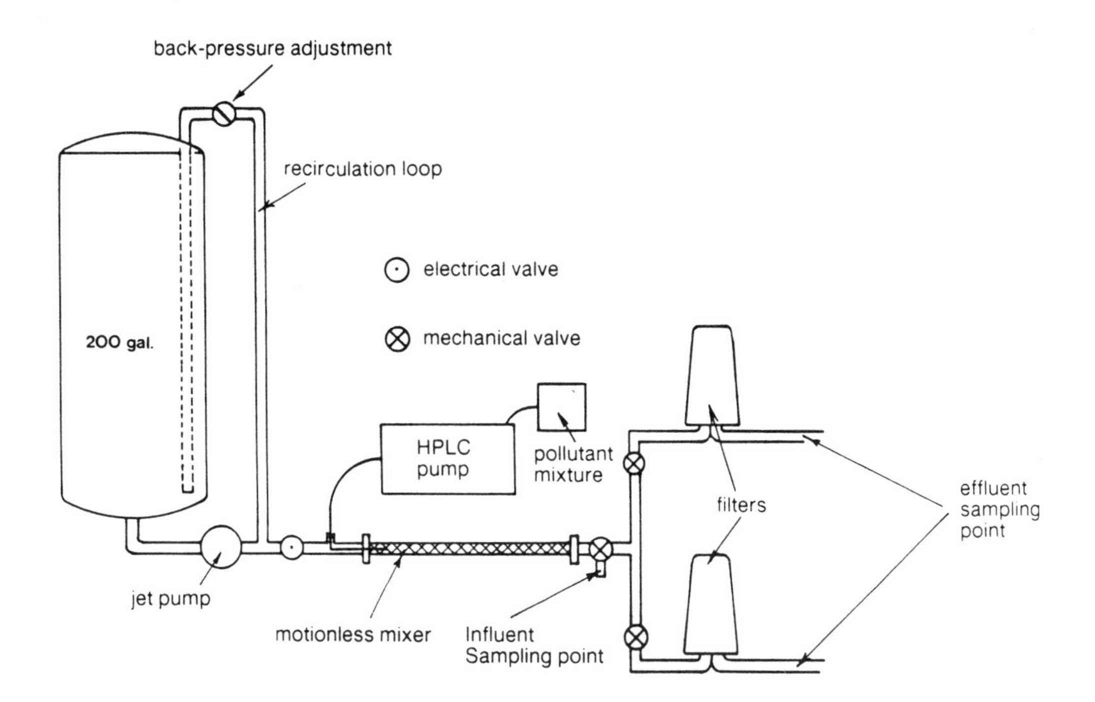

provides very consistent influent concentrations. The flow is controlled, influent and effluent samples collected for every sample point, and the temperature monitored. The test was run continuously for eight hours per day with an overnight stagnation period. Influent concentrations were generally chosen to be approximately 100 to 200 µg/l. Chloroform was higher to approximate the NSF standard requirement.

RESULTS

The results are contained in Tables 6 to 10. The trihalomethanes group shows only chloroform breaking through, at a concentration below 5 µg/l. The halogenated alkanes were subdivided for testing. The detection limit for 1,2-dibromoethane (EDB) was reduced to 5 ηg/l to accommodate states with limits of 20 ηg/l. The detection of trichlorofluoromethane and carbon tetrachloride are near detection limits. The halogenated alkenes include trichloroethylene and tetrachloroethylene, common pollutants. The aromatics group includes benzene, which based on Dobbs' isotherm data (8) would be expected to break through. Our data contradicts that expectation. The other aromatics are removed. The chlorinated hydrocarbons, including 1,4-dichlorobenzene are removed as would be expected based on efficacy for benzene.

These compounds provide a glimpse of the diversity of chemical compounds that can be removed by GAC. The difficulty of completing a project of this magnitude several times forced a good look at the trends. The potential use of a surrogate for most of the compounds tested appeared feasible although not yet adequately documented or accepted by the scientific or regulatory community.

GRANULAR ACTIVATED CARBON - PERFORMANCE

THEORETICAL CONSIDERATIONS

A brief look at the basic principles of GAC adsorption and some of the molecular properties affecting adsorbability of chemicals is helpful. Granular activated carbon has been demonstrated to adsorb a wide variety of organic chemicals. It has limited capacity for some classes of organic compounds, and for water soluble ions and metals. Carbon has been referred to as "black magic," but this is not really true. There is a good base of information on how GAC works (9,10,11). For a chemical to be adsorbed onto carbon, the attractive forces must be strong enough to overcome repelling forces. The forces of attraction are generally agreed to be primarily due to Van der Waals forces, which are relatively weak. First

Table 6. Class: Purgeables - Group: Trihalomethanes

Compound	Average Influent (µg/l)	Effluent @ 500 gal (µg/l)	Effluent @ 750 gal (µg/l)	Detection Limit (µg/l)	Total Loading (mg)
Chloroform	414	1.25	1.25	0.28	1,253
Bromodichloromethane	129	*	-	0.11	366
Dibromochloromethane	115	*	-	0.13	327
Bromform	204	*	-	0.93	580

Table 7. Class: Purgeables - Group: Halogenated Alkanes

Compound	Average Influent (µg/l)	Effluent @ 500 gal (µg/l)	Effluent @ 750 gal (µg/l)	Detection Limit (µg/l)	Total Loading (mg)
1,2 Dichloroethane	112	*	-	0.5	320
1,1-Dichloroethane	120	*	-	0.25	340
1,1,2-Trichloroethane	164	*	-	0.5	467
1,2-Dibromoethane (EDB)	59	*	-	0.005	169
1,1,1-Trichloroethane	116	*	-	0.16	329
1,2-Dichloropropane	118	*	-	0.34	334
Trichlorofluoromethane	112	*	0.8	0.51	318
1,1,2,2-Tetrachloroethane	151	*	-	0.31	430
1,2,3-Trichloropropane	146	*	-	2.7	413
Carbon Tetrachloride	78	*	0.2	0.22	222
1,2-Dibromo-3-Chloropropane (DBCP)	258	*	-	0.09	733

Table 8. Class: Purgeables - Group: Halogenated Alkenes

Compound	Average Influent (µg/l)	Effluent @ 500 gal (µg/l)	Effluent @ 750 gal (µg/l)	Detection Limit (µg/l)	Total Loading (mg)
1,2-Dichloroethylene	109	*	-	0.17	309
trans-1,2-Dichloroethene	96	*	-	0.11	272
trans-1,3-Dichloropropylene	170	*	-	0.14	482
Trichloroethylene	100	*	-	0.28	284
1,1,2,2-Tetrachloroethylene	92	*	-	0.27	262

Table 9. Class: Purgeables - Group: Aromatics

Compound	Average Influent (µg/l)	Effluent @ 500 gal (µg/l)	Effluent @ 750 gal (µg/l)	Detection Limit (µg/l)	Total Loading (mg)
Benzene	113	*	-	0.82	320
Toluene	114	*	-	1.0	325
Chlorobenzene	107	*	-	0.19	304
Xylene	292	*	-	3.9	830
Ethylbenzene	163	*	-	1.6	462

* Indicates value below detection limit (99 percent CI).

Table 10. Class: Base/Neutrals - Group: Chlorinated Hydrocarbons

Compound	Average Influent (µg/l)	Effluent @ 500 gal (µg/l)	Effluent @ 750 gal (µg/l)	Detection Limit (µg/l)	Total Loading (mg)
1,2 Dichlorobenzene	150	*	-	0.75	429
,3-Dichlorobenzene	184	*	-	0.49	526
1,4-Dichlorobenzene	120	*	-	1.05	342
Hexachloroethane	111	*	-	1.06	318
1,2,4-Trichlorobenzene	313	*	-	0.35	313
2-Chloronaphthalene	109	*	-	0.5	110
Hexachlorobutadiene	154	*	-	0.24	444
Hexachlorocyclopentadiene	58	*	-	1.03	171
Hexachlorobenzene	66	*	-	0.48	192

* Indicates value below detection limit (99 percent Cl).

the molecules must be brought close to the carbon surface. Particle size and design are important to minimize the water thickness around the carbon particles. Then the chemical must diffuse through the boundary layer to the carbon surface. It must diffuse into the pores, until it is tightly retained in the micropore region of the GAC. The kinetics of the surface diffusion or pore diffusion are limiting steps. Crittenden, Snoeyink, and Weber, among others, have developed theories and applied them to explain GAC behavior in actual use (12-15). For single solute models, predictions work quite well. When multiple solutes are being adsorbed on GAC, there is a competition for available sites, with the more strongly attracted molecules displacing the more weakly retained ones. These competitive effects are beginning to be understood, such that prediction of orders of desorption and capacity can be made when detailed information is available on the GAC being used. This is of great value to a manufacturer when designing POU devices and selecting a carbon source. It is not practical for evaluation of many different completed units due to the detailed information that must be developed for the GAC. However, the principles will be important in understanding results of dynamic testing.

It is important to note that GAC is not a generic commodity. The source of the carbon and how it is prepared for use provides a range of capacities, selectivities, and kinetics that must be evaluated.

Evaluation of GAC includes numerous tests. Surface area is important since adsorption is basically a surface phenomena. The surface area of carbon is very high, generally in the range of 500 to 1,500 m^2/g (2,445,500 to 7,327,600 sq ft/lb) mostly in internal pores which are like a maze of interconnecting channels. Pores below 20 angstroms (7.87 x 10^{-8} in) in radius, called micropores, are where most adsorption takes place with larger transition pores from 20 to 500 angstroms (7.87 x 10^{-8} to 1.97 x 10^{-6} in) providing access and some additional adsorption. Micropores over 500 angstroms (1.97 x 10^{-6} in) have little capacity for small molecules. The distribution of pore sizes and the percentage in the range of interest determine the actual performance of carbon. A BET nitrogen surface area measurement provides a great deal of information on surface area and pore size distribution. Particle size and particle size distribution are also important influences in the kinetics of adsorption.

Characterization of carbon pores includes traditional tests such as iodine values, generally agreed to measure small pores, with a pore radius of less than 20 angstroms (7.87 x 10^{-8} in). Carbons may have iodine values of 600 to 1,200. Molasses number, based on decolorization of molasses solutions, represents pore volume in the range of 20 to 500 angstroms (7.84 x 10^{-8} to 1.97 x 10^{-6} in). Dye adsorption, such as methylene blue, provides additional data on capacity for large molecules.

All of these measurements are useful for preliminary evaluation of GAC, but must be used with caution due to the multiplicity of interactions taking place under dynamic conditions.

APPLICABILITY

To determine whether GAC is applicable to removal of a chemical, it is necessary to look at some of the properties of the chemical that affect its adsorption. Removing a chemical from water is easier if it is not very soluble (Table 11). The solubility versus GAC capacity for several VOCs (capacities expressed as Freundlich adsorption isotherm values), shows a good correlation between low solubility and good adsorption. While the capacity in mg/g (lb/lb) of carbon may vary with carbon type, the relative order of elution is predictable.

Other factors, such as the substitution of bromine for chlorine increase adsorption, as evidenced for the THMs:

	Capacity X/MCo @ 100 μg/l
$CHCl_3$	3.05
$CHCl_2Br$	5.9
$CHClBr_2$	10.3
$CHBr_3$	17.0

Adding chlorine to a molecule increases adsorbability:

	Capacity X/MCo @ 100 μg/l
$CHCl_2$	0.45
$CHCl_3$	3.05
CCl_4	9.78
Benzene	13.47
1,4-Dichlorobenzene	196.4

Addition of double bonds to a molecule increases adsorbability:

	Capacity X/MCo @ 100 μg/l
1,2-Dichloroethane	2.88
trans1,2-Dichloroethane	8.32

As several of the factors are considered, a significant increase in capacity results:

	Structure	Capacity X/MCo @ 100 μg/l
Chloroform	CL - C(CL)(CL) - H	3.05

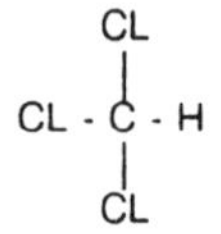

	Structure	Capacity X/MCo @ 100 μg/l
1,1,1-trichloroethane	CL - C(CL)(CL) - C(H)(H) - H	7.72

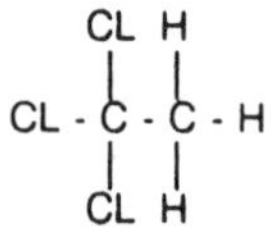

	Structure	Capacity X/MCo @ 100 μg/l
1,1,2-trichloroethylene	(CL)(CL) C = C (CL)(H)	25.17

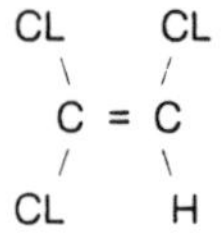

	Structure	Capacity X/MCo @ 100 μg/l
tetrachloroethylene	(CL)(CL) C = C (CL)(CL)	122.9

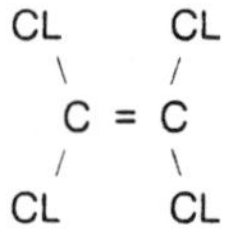

Table 11. Solubility vs. Capacity

Compound	Solubility @ 20°C (mg/l)	Capacity X/MCo @ 100 μg/l (mg/g)
Methylene Chloride	18,236	0.454
1,2-Dichloroethane	8,690	2.88
Chloroform	8,000	3.05
1,1,1-Trichloroethane	4,400	7.72
trans 1,2-Dichloroethylene	2,190	8.32
Benzene	1,780	13.47
Trichloroethylene	1,100	25.17
1,4-Dichlorobenzene	70	196.38

Based on capacity values obtained from Freundlich adsorption isotherms, it is obvious why chloroform or 1,1,1-trichloroethane would break through first, and why trichloroethylene and tetrachloroethylene did not break through in GSRI or field studies.

CAPACITY

Activated carbon has a finite capacity for any one compound. When multiple compounds occur, they compete to some extent for the available sites on the carbon, reducing the capacity for the less strongly adsorbed compound. Estimating capacity involves a number of considerations including:

- Which chemicals are present?
- What is the concentration of each?
- What is the maximum capacity of the carbon for the chemicals?
- What is the maximum effluent concentration allowed?

- What are the kinetics of the compound with the carbon?
- What are possible competing materials?
- How does the design of the POU device affect adsorption?
- How is capacity to be expressed?

CRITERIA TO DESIGN AND EVALUATE GAC-POU DEVICES

There are basically two main criteria that determine the efficacy of a GAC-based POU device:

- The capacity of the GAC used (isotherm capacities),
- The design of the final unit to approach maximum capacity (dynamic testing).

Capacity of the carbon is obviously critical. Performance cannot be achieved without a GAC that has a significant capacity for the chemicals of concern. One driving interest presently is the category of VOCs as proposed by EPA. This group is of special interest due to the difficulty of removal from water. It is a limiting factor in GAC performance. A GAC can be selected for a POU that provides a maximum adsorptive capacity for the smaller molecular weight VOC compounds. Generally, this comes at the expense of capacity for the higher molecular weight compounds, but this is not the limiting factor.

FREUNDLICH ADSORPTION ISOTHERMS

The capacity values used have been isotherm values that are maximum values at an effluent concentration of 100 µg/l. To discuss maximum theoretical capacity, it is necessary to bring in batch equilibrium experiments, referred to as Freundlich adsorption isotherms (11,16). Bottle point isotherms are run in a series of bottles containing different amounts of carbon, and a water solution of the chemical (e.g. 1 to 2 mg/l of VOC). The bottles are sealed and mixed for seven days, the carbon filtered out, and the concentration remaining in the water measured. The carbon weights and VOC concentrations must be adjusted to yield final solution concentration of 1 to 200 µg/l. This allows calculation of the quantity adsorbed on the carbon, expressed as mg/g carbon.

The Freundlich equation is expressed as:

$$X/M = KC^{1/N}$$

Plotted as a straight line:

$$\text{Log}(X/M) = \text{Log } K + (1/N) \text{ Log } C$$

where,

X/M = the amount of component adsorbed (mg/l) divided by the weight of carbon (g/l)

C = the equilibrium concentration of the component (mg/l or µg/l)

1/N = slope of the line for the component in solution

K = Constant for each compound: $(mg/g) \times (l/mg)^{1/N}$ or $(\mu g/g) \times (l/\mu g)^{1/N}$

It must also be stated that to compare isotherm behavior, the chemicals must be grouped such that the similar molecules are compared. The adsorption isotherm curve, when plotted as the logarithm of both sides of the equation, yields a straight line providing significant information (Figure 3). One can:

- Screen potential activated carbon samples for use.
- Provide an estimate of the adsorption capacity of a component by activated carbon.
- Determine if the desired effluent can be obtained with a given amount of activated carbon.
- Estimate the capacity difference when equilibrium concentrations are changed.
- Predict the relative breakthrough order of adsorbates during column studies by comparing the capacity values obtained from isotherms.

From the isotherm of chloroform, one can find the capacity of that carbon at an equilibrium concentration of 100 µg/l, the current MCL for total trihalomethanes. This carbon could hold 3.05 mg $CHCl_3$/g carbon (0.003 lb $CHCl_3$/lb carbon). One can also readily see the impact of decreasing the equilibrium concentration to 25 µg/l; capacity is now 1.29 mg/g GAC (0.0013 lb/lb GAC). Reducing it further to 5 µg/l, the MCL concentration for most of the VOCs, the capacity is now down to 0.47 mg/g GAC (0.0005 lb/lb GAC). These values may change with different carbons or different operating conditions.

Absolute values for adsorption isotherms for the same chemical can show significant variations between published data, which may be due to any one or a combination of factors including the carbon used, pore size distribution, particle size, the time to equilibrium, the water temperature, pH, analytical techniques, and the number of points used to determine the line, among other factors. One factor is the presence of hardness ions and TOC. The effects of deionized water versus a municipal water were investigated. No significant difference is seen,

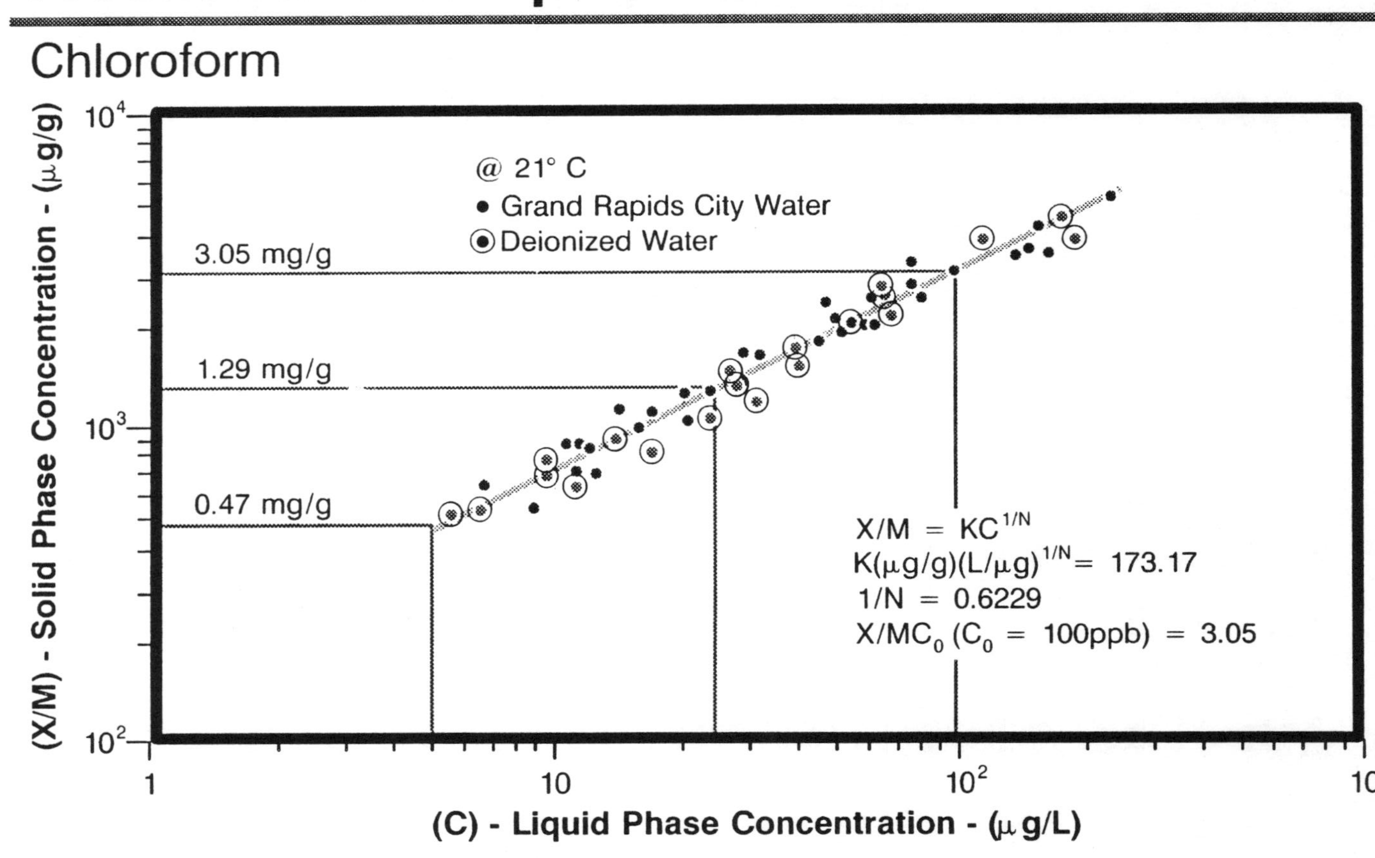

Figure 3. Freundlich adsorption isotherm.

however, the effects of temperature make a dramatic difference in capacity (Figure 4). As the water gets colder, the capacity of the carbon increases. Temperature control, or at least measurement, is an important element of capacity evaluation.

RELATIVE ISOTHERM VALUES

The real value in using adsorption isotherms for predicting performance of a POU device is in using the relative capacities for related compounds. By plotting the isotherms for related compounds (Figure 5), one can readily see that a unit that removes chloroform to a concentration of 5 µg/l will also remove 1,1,1-trichloroethane, trichloroethylene, and 1,4-dichlorobenzene. While the total quantity of compound absorbed may differ from unit to unit, the relative capacities, or the order of breakthrough of the compounds will remain the same. These general observations were made during the Gulf South Research Institute study, and also during the NSF/EPA studies at Silverdale and Rockaway. A review of the literature on isotherms supports the relative ranking, although absolute values differ (8,9,11,17,18).

Therefore, if one compound is chosen as a model, or surrogate compound, adsorption for the other compounds is rather certain. By selecting the concentration of the model compound, e.g. chloroform, at the highest level anticipated for a known contaminant, removal is assured for at least the rated capacity of chloroform, and longer if the compound is more tightly retained.

DYNAMIC TESTING AND ISOTHERM CORRELATION

The gap between theory, that is maximum capacity as determined by isotherm, and practice is a measure of how well a POU device has been designed. This is done by dynamic testing. The isotherm capacity is determined on the GAC under near ideal conditions. The actual performance of the unit takes all factors into consideration. The contact time between water and carbon in POU devices is typically short, in the order of seconds rather than minutes for large scale GAC contactors. Kinetics become a factor in selecting the carbon source. Since the active sites in GAC are internal and the pores very small, it takes time for molecules to diffuse into them. Design must take into account the pore size distribution, particle size, and particle size distribution. As particle size gets smaller, adsorption gets faster but backpressure gets higher and flow rate drops. This may be a limiting factor. The chromatographic effect, where strongly adsorbed materials are retained at the early portion of the bed, reduces competition so only similar compounds of similar capacity compete for pores. The mass transfer zone is created where the first GAC contacted is loaded to its capacity first and the concentration decreases through the bed until the compound is completely removed. This wavefront, or mass transfer zone moves through the carbon bed until the compound elutes. Either a long bed, or a short mass transfer zone is needed for optimum performance. Again, this is an important criteria when designing a unit to remove VOCs to below 5 µg/l.

Results of dynamic tests are shown for chloroform well past the point of breakthrough (Figure 6). It illustrates that a lot of capacity may be left after breakthrough reaches 5 µg/l. The typical single solute chloroform curve with an average influent of 108 µg/l is shown plus a typical chloroform curve at an average influent of 442 µg/l. The capacity decreases significantly as the influent concentration is raised. The multisolute curve for chloroform at 102 µg/l shows a reduction in capacity but not as severely as the 442 µg/l curve. The point of breakthrough is sooner as the load on the filter increases, but the difference is less at the low µg/l concentration. This demonstrates that competition can be accounted for by increased challenge concentrations, and the multisolute curve also demonstrates a limited competition since the total challenge greatly exceeded the 442 µg/l level.

The effects of multisolute interactions can be seen in Figure 7. Here, the chloroform curve can be seen in relation to the other VOCs as part of a 14 component run. Seven of the VOCs are plotted here. This illustrates the predictive power of relative isotherms since the components of the multicomponent mixture elute in the same relative order as that shown by the isotherms summarized in Table 12.

In these cases, testing capacity of the POU device for chloroform would represent a worst case. If chloroform is removed, so too is every other compound. If $CHCl_3$ testing for capacity is done at a level equalling the concentration of any single component it could provide a minimum capacity of the POU device.

FURTHER EVALUATION TECHNIQUES

The data from dynamic studies provides practical data, but limited to the specific chemicals tested. The use of isotherm data to predict elution orders lends support to the surrogate concept. There are other theoretical treatments that lend further confirmation that the practical behavior complies with theory. The use of Polanyi liquid phase adsorption potential theory deals with the energy involved in the adsorption process (10). From limited isotherm data it is possible to predict capacity for other related compounds. Further theoretical modelling based on complex computer programs can provide predicted performance for single solute or multisolute situations to compare to actual dynamic data. Successful prediction of performance for several of the VOCs confirms that the basic theoretical treatments are correct. Use of minicolumns to predict the performance of full scale operations has great advantages for large scale operations (19,20). It also

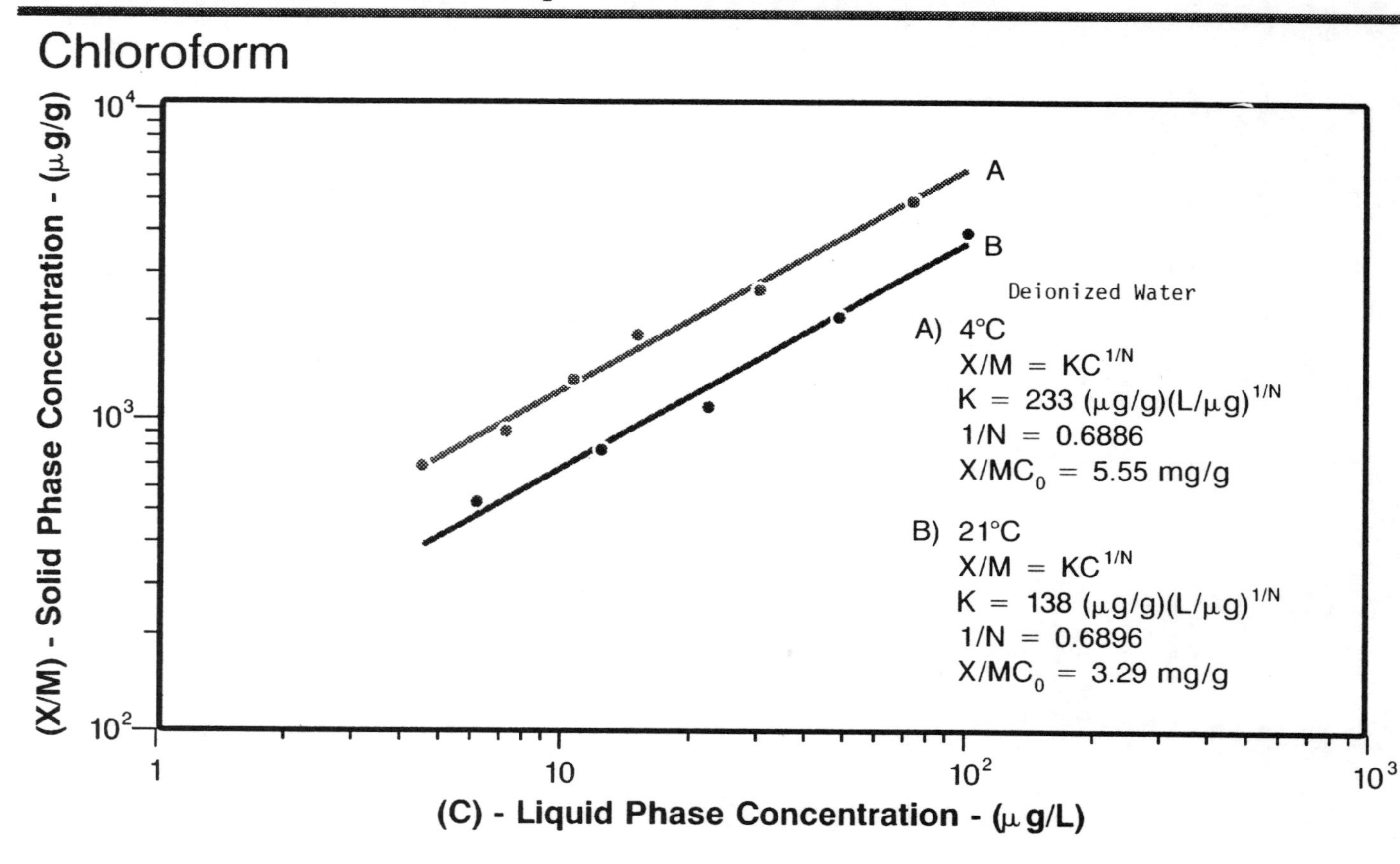

Figure 4. Freundlich adsorption isotherm.

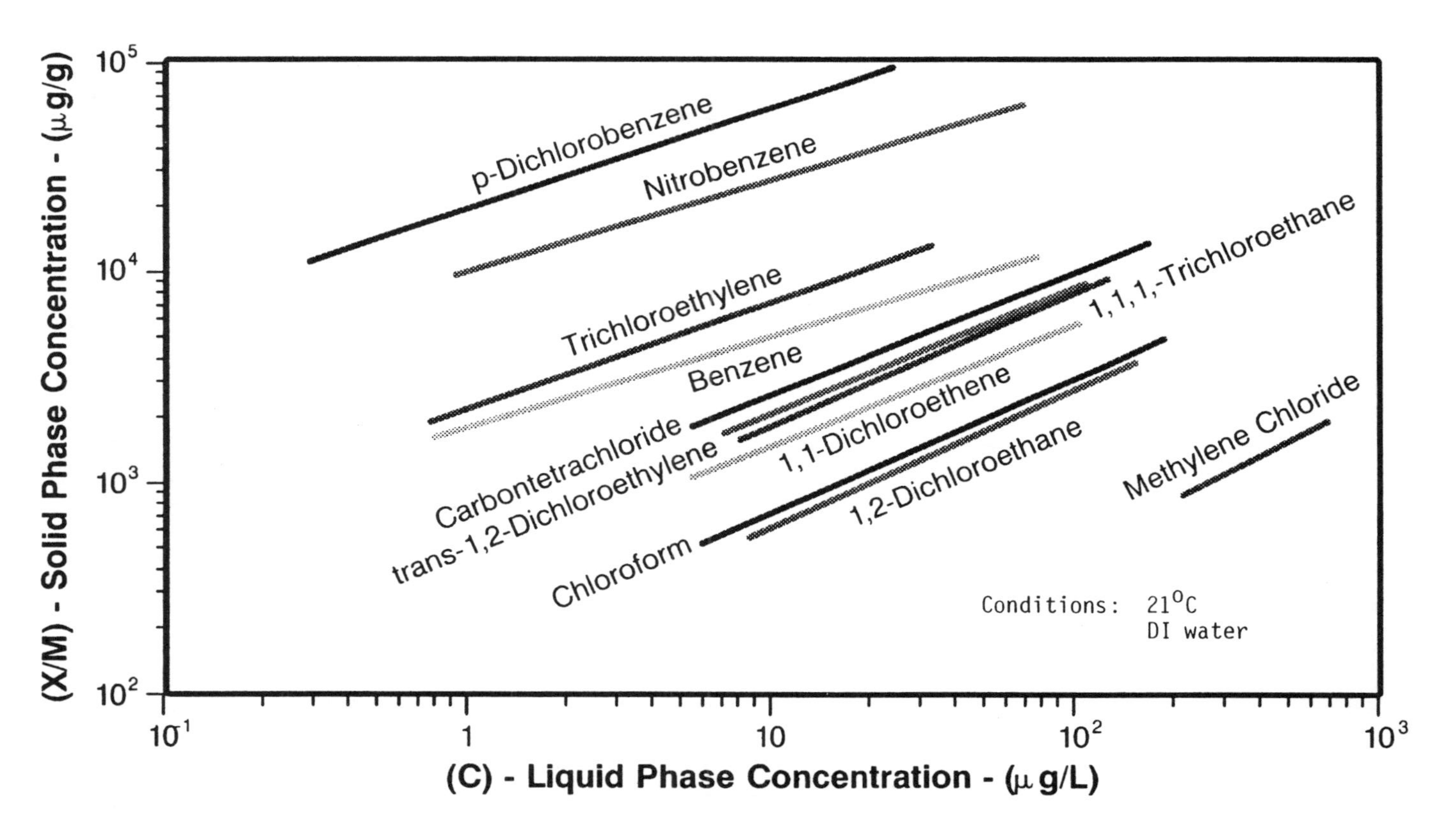

Figure 5. Freundlich adsorption isotherms.

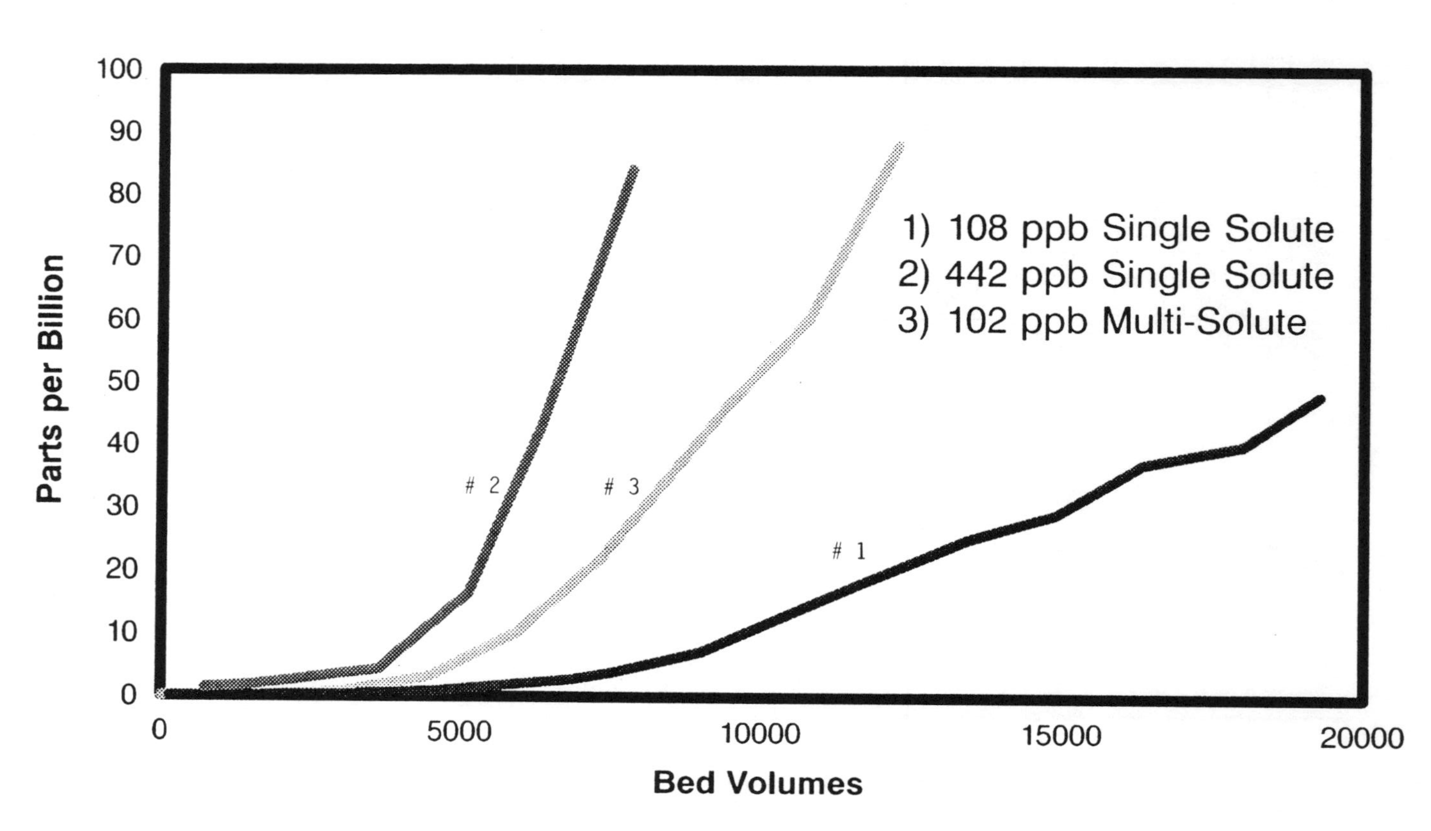

Figure 6. Chloroform dynamic breakthrough curves.

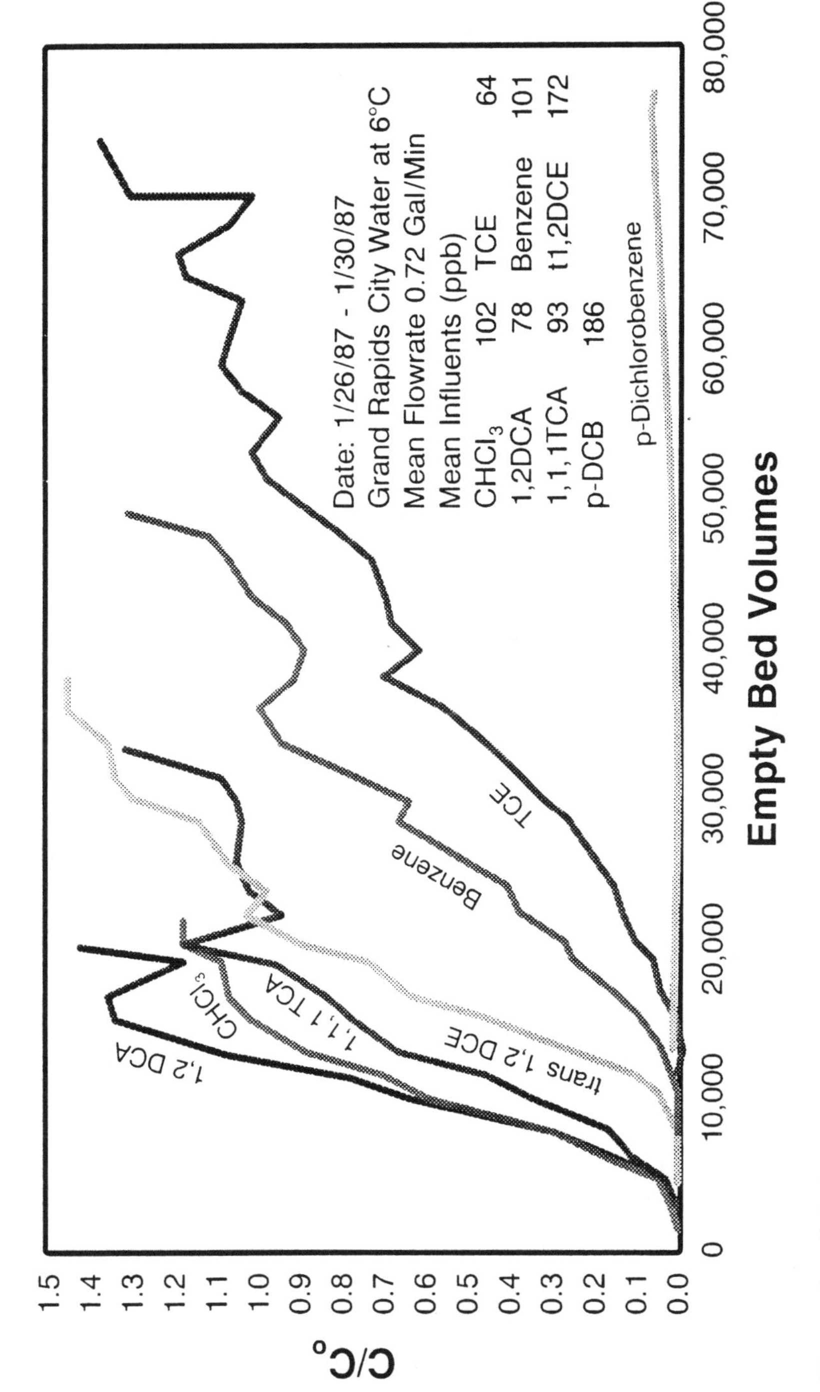

Figure 7. Multi-component study.

Table 12. Isotherm Capacities

Compound	Capacity X/MCo @ 100 µg/l (mg/g)
1,2-Dichloroethane	2.88
Chloroform	3.05
1,1,1-Trichloroethane	7.72
trans 1,2-Dichloroethylene	8.32
Benzene	13.47
Trichloroethylene	25.17
p-Dichlorobenzene	196.38

can provide valuable information on carbon performance in much shorter time than full scale testing. Further, detailed information on the carbon is available, and must be obtained for some of the predictive modelling programs.

While these more involved evaluation methods may be too complex for unit evaluation on a regular basis, they do lend support to the dynamic performance data and the surrogate approach supported by relative isotherm approach.

CONCLUSIONS

What can one conclude from this for evaluation of POU devices?

- A properly designed POU can be effective in removing the regulated VOCs to below the MCL concentrations for a significant period.
- The removal of the THMs, in particular chloroform, provides assurance that the other VOCs can be removed as well.
- Isotherm capacities can predict relative breakthrough orders for the VOCs.
- Testing with a surrogate, e.g., chloroform, provides a conservative capacity value for the other VOCs.
- Temperature is an important parameter in POU testing.
- Selection of GAC and unit design is important.

Results from lab and field studies have demonstrated the chemical removal ability of POU devices. Theoretical projections can be valid to predict approximate lifetimes, and breakthrough orders of compounds. The concept of a surrogate chemical for capacity testing is valid, if precautions are taken in selecting the chemicals. There is a significant body of research already supporting the surrogate concept if properly evaluated.

ACKNOWLEDGEMENTS

We would like to acknowledge John Wezeman for his work performing the VOC analyses, Tim Modert for the isotherms, Karen Vanderkooi for the data reduction and plotting, and Roy Taylor for his expertise in making units for testing.

REFERENCES

1. Safe Drinking Water Act, (42 U.S.C. 300(f), et seq). Ammended June 19, 1986.

2. National Primary Drinking Water Regulations: Synthetic Organic Chemicals; Monitoring for Unregulated Contaminants. Fed. Reg. 52, 130:25690, July 8, 1987.

3. Development of basic data and knowledge regarding organic removal capabilities of commercially available home water treatment units utilizing activated carbon. Gulf South Research Institute Reports to the Office of Drinking Water. USEPA, Phases 1-3, NTIS PB 82-159-583, Springfield, VA, May 1979 - October 1981.

4. Point-of-use reduction of volatile halogenated organics in drinking water. National Sanitation Foundation Report to Water Engineering Research Lab, U.S. EPA, 1985.

5. Drinking Water Treatment Units, Standard 42 and Standard 53. National Sanitation Foundation, June 1982.

6. Van Dyke, K. et al. The standard design and testing for a pressed carbon block water filter. American Laboratory. 18(a):118, September 1986.

7. The Amway water treatment system: distributor/customer information packet. Amway Corporation, Ada, MI, 1987.

8. Dobbs, R.A. and Cohen, J.M. Carbon adsorption isotherms for toxic organics. EPA-600/8-80-023, U.S. EPA, Cincinnati, OH, 1980.

9. Perrich, J.R. Activated carbon adsorption for wastewater treatment. CRC Press Inc., Boca Raton, FL, 1981.

10. Suffet, I.H., and McGuire, M.J. Activated carbon adsorption of organics from the aqueous phase, Vol 1. Ann Arbor Science, Ann Arbor, MI, 1980.

11. Crittenden, J.C., Hand, D.W., Arora, H. and Lykins, B.W. Design considerations for GAC treatment of organic chemicals. Jour. AWWA. 79:1:74, Jan. 1987.

12. Crittenden, J.C., Wong, B.W.C., Thacker, W.E., Snoeyink, V.L. and Hinricks, R.L. Mathematical model of sequential loading in fixed bed adsorbers. Jour WPCF. 52:11:2780, Nov. 1980.

13. Thacker, W.E., Crittenden, J.C. and Snoeyink, V.L. Modelling of adsorber performance: variable influent concentration and comparison of adsorbents. Jour WPLCF. 56:3:243, March 1984.

14. Crittenden, J.C. and Weber, W.J. A model for design of multicomponent adsorption systems. J. Environ Eng. Div., Am. Soc. Civil Eng. 104(EE6):1175, 1978.

15. Kuang-Tasn, L. and Weber, W.L. Characterization of mass transfer parameters for adsorber modelling and design. Jour WPCF. 53:10:1541, Oct. 1981.

16. Randtke, S.J. and Snoeyink, V.L. Evaluating GAC adsorptive capacity. Jour AWWA. 75:8:406, Aug. 1983.

17. Sontheimer, H., Frick, B., Fettig, J., Horner, G., Hubele, G. and Zimmer, G. Adsorptionsverfahren zur Wassereinigung, DVGW-Forschungsstelle am Engler-Bunte-Institut der Universitat Karlsruhe (TH), 1985.

18. Crittenden, J.C., Luft, P. and Hand, D.W. Prediction of multicomponent adsorption equilibria in background mixtures of unknown composition. Water Research. 19:12:1537, 1985.

19. Crittenden, J.C., Berrigan, J.K. and Hand, D.W. Design of rapid small-scale adsorption tests for a constant diffusivity. Jour WPCF. 58:4:312, Apr. 1986.

20. McGuire, M.J. and Suffet, I.H. Advances in Chemistry Series, No. 202. Treatment of water by granular activated carbon. American Chemical Society, Washington DC, 1983.

PERFORMANCE AND APPLICATION OF RO SYSTEMS

Donald T. Bray
Desalination Systems, Inc.
Escondido, CA 92025

BASIC CONCEPTS

Figure 1 shows some basic concepts related to reverse osmosis (RO). On the top of the figure there is a scale in angstroms (A) and µm. These are referenced to the pore size ranges of the membrane-filtration business. An angstrom is about the size of a hydrogen atom (10^{-8} cm [3.87 x 10^{-9} in]) and a water molecule is around 2 A (7.9 x 10^{-9} in). The RO field is at the extreme lower end of the pore size range, and covers the range 1 to 5 A (3.94 x 10^{-9} to 2 x 10^{-8} in). Ultrafiltration (UF) covers the range from about 10 A to about 1,000 A (3.9 x 10^{-8} to 3.9 x 10^{-6} in) micro filtration from 0.1 to 1 µm (3.9 x 10^{-6} to 3.9 x 10^{-5} in) and general filtration above about 1 µm (3.9 x 10^{-5} in). Other people may use slightly different divisions. My subsequent discussions will be concerned with the RO range.

Also shown on Figure 1 is an important concept: the basic difference between RO/UF devices and filtration devices. RO/UF devices are separative devices; they take a feed stream and separate it into two parts -- a product and a reject. There is no accumulation within the unit. It can operate continually without buildup. Filters, on the other hand are accumulators. Particles are removed from the feed and accumulate in the unit itself. Hence, the filters have a finite life and the characteristics of the product are continually changing. A shut-off provision after a given throughput is being considered for accumulator type devices. Such a concept does not apply to separative devices such as RO, and a different safeguard approach needs to be taken here.

Figure 2 shows a highly stylized rendering of the surface of cellulose acetate (CA). Note the long roundish particles of CA with various bumps protruding. Actually the CA molecules are much longer relative to their diameter and of course are more twisted and intertwined. One might represent thin film membranes (TFM) in somewhat the same fashion but more crooked and with occasional cross links. The spacing between molecules might vary from 0 to 5 A (0 to 2 x 10^{-8} in). The molecules are in rapid motion, vibrating several thousand times a second so the spacings are continually changing. Hence, we have no pore size per se in RO. Only when we get to UF do we start having definable pore sizes. Also shown is a water molecule. It is moving very fast, making thousands of collisions per second, and under influence of a driving force (in this case pressure), can enter the spacings between the CA molecules and diffuse through the network essentially one at a time. Also shown is a Na^+ ion surrounded by a group of water molecules. Water molecules, being dipolar, will attach themselves to the $^+$ charged surface of the Na^+ ion forming a molecule several times as large. It will move through the lattice in the same way as water. However, it can't move nearly as readily as water since it is much larger. Hence, there is, in effect, a separation of Na^+ and water molecules with the water molecules going through the membrane and the Na^+ accumulating on the surface until back diffusion through the laminar layer to the reject stream removes it from the unit. One can imagine that a virus would look like a huge boulder on the surface. One might also note that the separation efficiency of the membrane will vary with the shape of the molecule -- e.g., whether rod-, sheet-, or ball-shaped. Basically, however, the larger the molecule, the better the rejection. The above is a simplistic picture. Surface effects such as the electron structure of both membrane and diffusing molecule also come into play, but the physical size approach is a good first approximation.

TYPES OF SYSTEMS

Figure 3 lists the three types of RO point-of-use (POU) systems in use today. I should note that all use a spiral-wound membrane assembly configuration. The first commercial unit on the market was an over-the-counter (OTC) type made and marketed by Culligan in 1965 to 1966. It hung on the wall of the kitchen as shown schematically in Figure 4. The spiral-wound element and pressure vessel were located inside the product storage tank. Note locations of feed, reject, overflow, and product. This unit had several drawbacks and was never very successful due in part to installation difficulties of feed, reject, and especially the gravity overflow.

Figure 1. Reverse osmosmis - basic considerations.

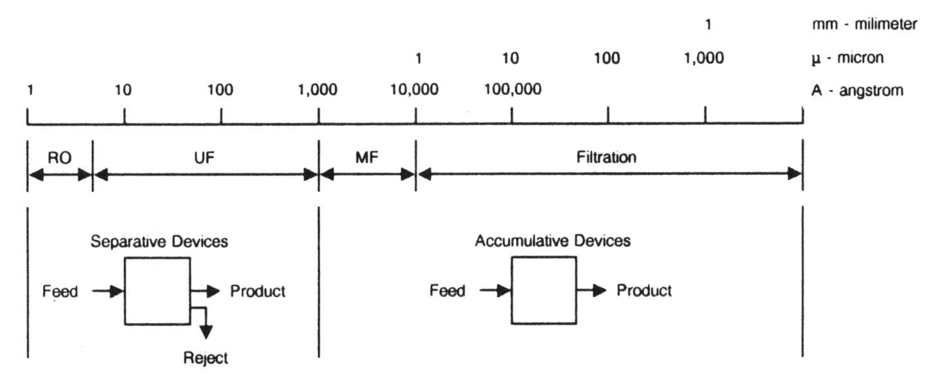

Figure 2. CA membrane surface schematic.

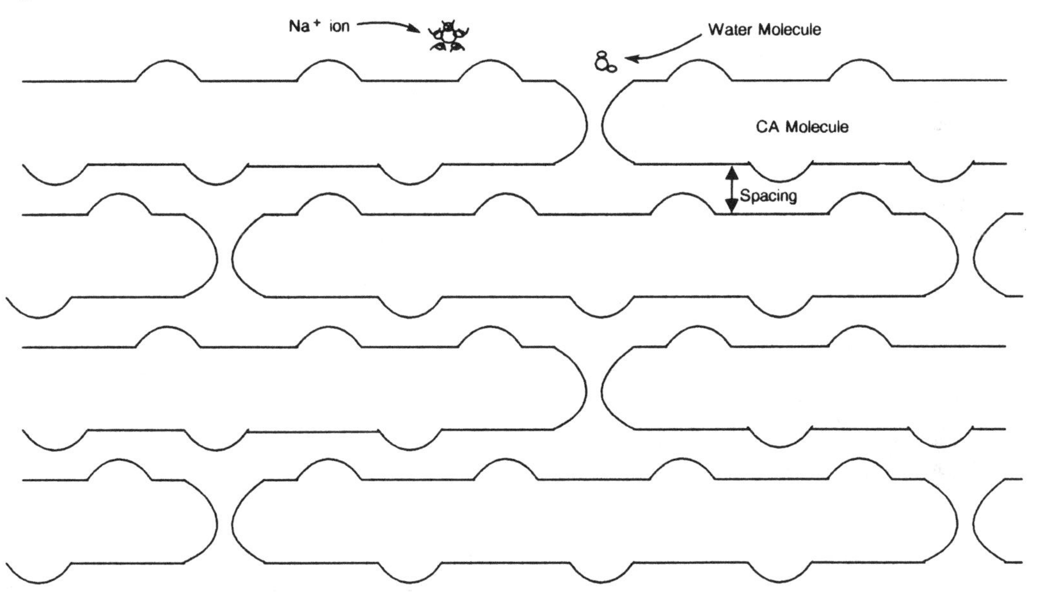

These early units had no post carbon filters so there were some taste problems. Also, the gravity flow of product from the spigot was replaced by air in the top of the storage tank which introduced dust particles with each use, so the heterotrophic bacteria species within the storage tank kept changing. Further, and in hindsight, the marketing and selling techniques of RO POU still needed to be developed.

The next generation of OTC units were counter-top models that sat on the counter with feed connections to the sink faucet, and drained directly into the sink.

Figure 3. Types of RO/POU systems.

UTC	-	Under the Counter
OTC	-	Over the Counter
OU	-	Office Unit

This eliminated the installation and overflow problems of the original, wall mounted unit but resulted in use of valuable counter space and retained the continued taste and contamination problem. The third generation of OTC, as shown in Figure 5 and made by Nimbus

Figure 4. OTC concept.

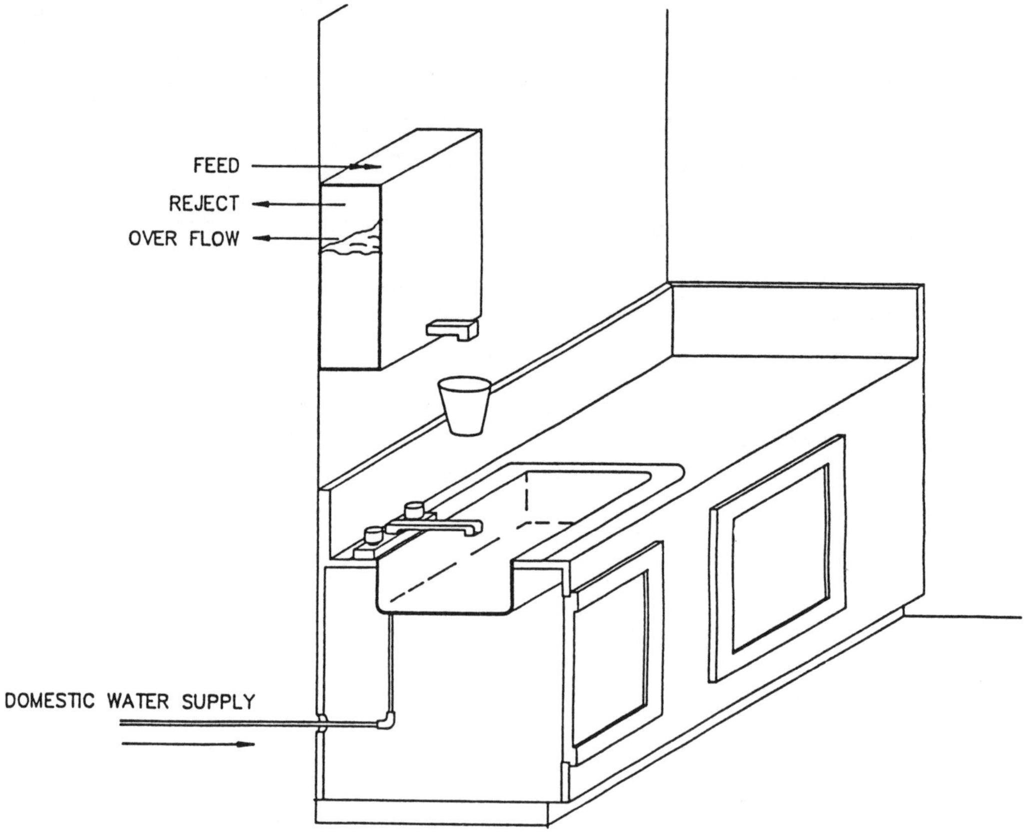

Water Systems, is of more recent vintage in which the unit is attached directly to the faucet with a quick disconnect. Reject goes to the sink; product to a collection bottle. The unit is hooked up at night, removed in the morning and stored in the refrigerator, as is the product water container. It overcame most of the difficulties encountered in the other two approaches but adds the inconvenience of hookup and removal.

The second type of RO POU system developed was the office unit (OU), the first of which appeared in 1966. It was really an adaptation of the wall-mounted unit but utilizing the olla, or water storage compartment, of the conventional bottle water stands. Most of the original limitations of the wall-mounted unit still applied, but were offset to some extent by increased revenues per unit (to compensate for installation costs), and use of a product level operated shut-off valve and post carbon filter. More recently, there has been a trend toward use of pressurized product storage tanks, as used in the under-the-counter (UTC) systems.

The UTC was developed in 1966 and 1967 in an effort to overcome the limitations of the over-the-counter and office units. Figure 6 shows the basics of a UTC unit. All the components except the faucet are located under-the-sink. Tap water feed under pressure enters a pressure vessel containing a spiral-wound membrane element. The reject flows through a small capillary tube to drain. The pressure is reduced from line pressure at entrance to discharge pressure at outlet over the length of the capillary. The amount of reject is determined by the inside diameter and length of the capillary tubing. It is generally set at four to five times product flow rate. This ratio was chosen based on field experience. Basically, one can only take out that percentage of water until saturation of the least soluble component occurs, which in San Diego, is calcium carbonate at about 20 percent water removal. The product flows

Figure 5. Mint II - hang-on unit.

out of element into a sealed storage tank at a continuous but slow rate (10 to 20 ml/min [0.34 to 0.68 oz/min]). The tank has a rubber diaphragm with a low air pressure on one side (35 to 55 kPa [5 to 8 psi] when tank contains no product water). As the product flows into the tank, it moves the rubber diaphragm down, compressing the air. When the faucet is opened the water is forced out of the tank by the compressed air to point-of-use.

Some additional accessories need to be added to make a working sytem, as shown in Figure 7. A post carbon unit is placed in the product line between the storage tank and faucet. It serves as a final polisher to remove any tastes which were picked up from the tank material or that came through the membrane. There needs to be either a pressure relief valve on product side or a shut-off valve on the feed line operated by tank pressure. If a relief valve is used, it is generally set at about 1/2 line pressure. When the tank pressure reaches this point it opens the relief valve allowing the product to join the reject line. The unit continues to operate producing good water -- it just goes down the drain. When water is drawn from the storage tank and the pressure is reduced, the valve closes and good quality product refills the tank.

Figure 6. UTC - basics.

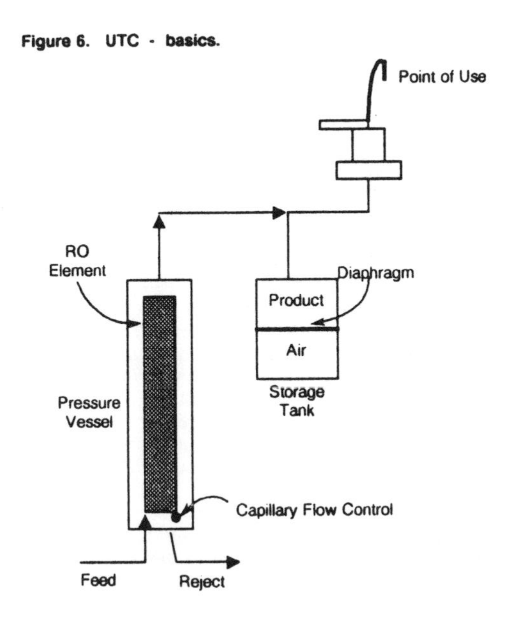

Figure 7. UTC - complete system.

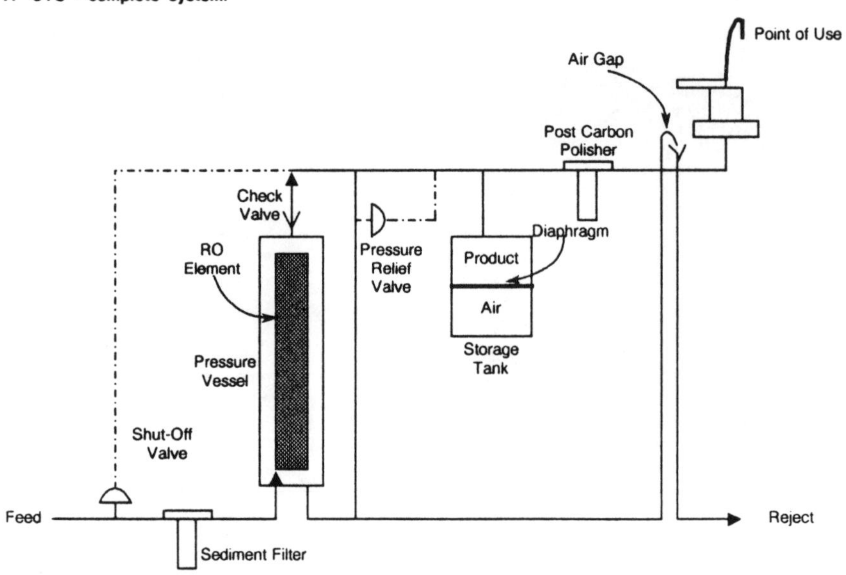

If a shut-off valve is used, it is located in the feed line, is operated by tank pressure, and is set at 1/2 to 2/3 line pressure. In many locations, a sediment filter (25 µm) needs to be added. When a TFM is used, a precarbon filter is also added to the feed line after the sediment filter to removal chlorine (which damages the polyamide TFM). The reject is shown flowing to an air gap above the sink. This is needed to meet plumbing codes requiring a positive break between the potable water and the sewer -- i.e., an air gap. The above UTC concept was developed in 1967 and is little changed to this day. Essentially, all UTC units made and in use today, utilize this concept.

USE DATA

Figure 8 shows my estimates of some key use data. There are about 500,000 units in use today. Total 1987 income to manufacturers was between $15 and 25 million from new units and about $20 million from replacement parts and elements, for a total of about $40 million per year. The total income to dealers and distributors was $50 to 75 million from new placements and perhaps $20 million from rental or leased units for a total of about $100 million per year. The cost to consumers to rent or buy is also given in the figure. Note that the little Mini II hang-on competes very well in production and quality, at a much lower price. It was developed to bring lower cost, good quality drinking water to the lower income and elderly as replacement for store or vending machine-purchased drinking water. To a large extent, it has succeeded in doing so.

MEMBRANE TYPE

There are two types of membranes on the market today that account for essentially all the membranes used in RO POU units. These are the CA types and the TFM polyamide types. I have grouped all the cellulosic base membranes into "CA Types." This includes cellulose acetate, diacetate, blend, triacetate, and cellulosic esters. There are only minor differences in their performance or limitations. Most of the units in the field today use CA, but the percentage of TFM is increasing and perhaps half of the new units placed in 1987 were TFM. Figure 9 lists the characteristics of the two types.

BACTERIA - VIRUS

No discussion of RO POU would be complete without a comment on bacteria/virus. When referring to Figure 2, I noted that viruses are huge compared to the spacings between molecules. When the membrane is sound there is no leakage. Membranes today can be made virtually defect-free. However, current POU systems cannot be sold as removing all bacteria and viruses for a few reasons. First, bypass leakage exists in all elements including the spiral-wound type. This leakage historically has been on the order of 0.1 to 0.5 percent; i.e., 0.1 to 0.5 percent of the product does not go through the membrane but through nonmembrane locations such as the glue lines, leaks, glue area weepage, O-ring leaks, etc. Secondly, there is no readily available fast response fail-safe systems such as a conductivity meter that tells us

Figure 8. Miscellaneous data.

Item	UTC	OTC	OU
Units Placed in 1987	60,000-100,000	25,000-50,000	5,000-10,000
Total Units in Use, end 1987	300,000-400,000	70,000-100,000	30,000-50,000
Manufacturer's Selling Price, $	150-250	30-70	350-450
Cost to Consumer, $ to buy	500-700	70-150	NA
$/month to rent	16-20	NA	25-40
Production, gpd	3-10	3-6	6-15
General Ionic, percent CR	80-95	85-97	80-95
General Inorganic Reduction	Varies	Varies	Varies

Figure 9 Membrane types - comparison of merits.

	CA Types	TFM Polyamide Types
Advantages	More experience Low cost Good ionic rejection	High flows High rejections Wide pH range (4-11) High temperature (120°F) Bacteria resistant
Disadvantages	pH range limited to 3-8 Temperature limited (≤100°F) Subject to bacteria attack Only fair organic rejection	More expensive Less experience Needs carbon prefilter Tendency to foul

Figure 10. Nimbus N-3A - a 20-year service call record.

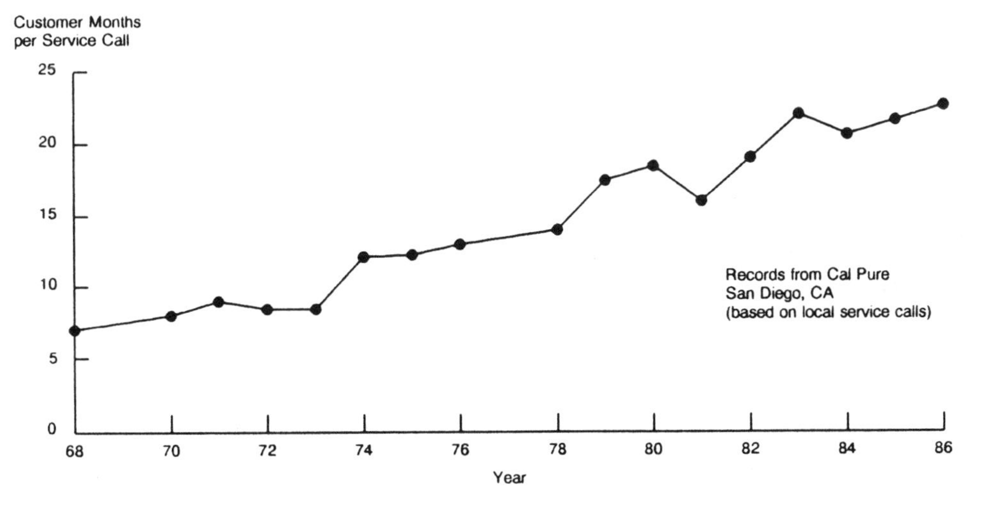

when a unit fails as regards bacteria/virus removal. Therefore, for these two reasons, the RO POU cannot warrant use on a nonpotable water supply. On the other hand, if I lived on a farm and had no readily available potable source, I would surely use an RO POU unit.

As shown in Figure 7, bacteria are always present in the product part of the system. They are introduced during installation from dust particles in the air, from handling, etc. They will build up to a semi steady state population depending on amount of use, temperature, species, amount of organics present, etc. Dr. Lee Rozelle, in his article in this publication, refer to the July 1987 standard setting a heterotrophic plate count of 500/ml or less. This number was related to the ability to accurately test for *E. coli* and not the health aspects of heterotrophic bacteria in the drinking water. Several years ago, I tested several hundred RO POU systems in the San Diego area. I also tested bottled water cabinet model drinking units. Essentially, none complied with this criteria. If this regulation were to be enforced across the board today, it would destroy the RO POU industry. In fact, the only unit on the market today that would meet this criteria is the nonstorage system; i.e., the hang-on faucet unit. I would suggest a look at developing different analytical techniques for determination of *E. coli* in the presence of other bacteria, if this is the limitation.

Progress has been made over the years in the technical service area. I would like to share with you the record of one dealer -- Cal Pure in San Diego. Figure 10 shows the service call record for the last 20 years for the Nimbus N-3A model. This data is for several thousand rental units. I use customer-months per service call as the criteria of excellence. This is the average number of months a customer's unit is in operation between all service calls regardless of nature of the call. The original design objective was 12 customer months per service call. As you can see, it started at around six and gradually increased to a current level of about 22. If the unit is owned by the customer rather than rented, the customer months per service call tend to be several months longer. This is a remarkable performance record for a unit that is generally installed by unskilled labor, operates unattended for long periods of time, retains a constant reject flow of very small magnitude, and has check valves that seal under "1" water pressure over these time periods. I might note that the dips in the curve were the result of minor system design changes. One must be very careful of design changes in these systems. If we plotted the cartridge life in San Diego instead of customer months against years, it would add about 10 months to the vertical part of this curve.

CONCLUSIONS

- RO POU is a relatively old concept -- going on 20 years. It has been slow in developing but has picked up speed in the last few years. There are around 500,000 units in use in the U.S. today.

- The economic value in 1987 was about $40 million to the system manufacturers and $100 million to the dealers and distributors.

- The challenge of the immediate future in the nonhealth-related areas for RO POU is marketing, sales, and service. We have satisfactory membranes to cover the needs, and the technical aspects are sufficiently in hand.

PERFORMANCE AND APPLICATION OF ULTRAVIOLET LIGHT SYSTEMS

Clyde Foust
Ideal Horizons Inc.
Rutland, VT 05701

INTRODUCTION

The increasing demand for high quality water, combined with the reduction in quality of available water sources, has caused an upsurge in water treatment technology. Ultraviolet (UV) light as a disinfectant was first discovered in 1847 and first commercially applied in 1901. The effectiveness of UV as a preferential microbiological treatment should be evident in its choice as the disinfectant in over 2,000 cities in Europe. Ultraviolet light accomplishes this without using harmful chemicals, some of which when used can lead to the formation of harmful secondary chemical products, such as trihalomethanes. Ultraviolet light does not impart any taste or odor nor is it possible to overtreat water using this method. Properly designed units cannot expose users to any harmful products. The effect of no measurable residual disinfectant in the water has been well studied in Europe and found to be of no concern. Maintenance of these systems is usually limited to annual servicing.

Most of the commercially available units use low pressure mercury vapor lamps. Operationally and mechanically, these lamps are very similar to the fluorescent lighting with which we are all familiar. Most units are quite simple in principle because the mercury vapor bulb in conjunction with a pressure reaction chamber is all that is required. However, in application, the units are considerably more complicated. The commercially available units must also be fitted with monitors, flow controllers, and other operational equipment. To treat water with ultraviolet light, the three parameters necessary to consider are the organism, dosage, and unit design.

ORGANISMS

The UV spectrum in this article refers to the spectrum between 200 and 300 nm (7.87 x 10^{-6} and 1.18 x 10^{-5} in). As a point of reference, the entire visible spectrum is 380 to 700 nm (1.5 x 10^{-5} in to 2.76 x 10^{-5} in) with the major spectrals between 450 and 600 nm (1.77 x 10^{-5} in and 2.36 x 10^{-5} in) The inactivation of organisms occurs through photophysical damage imparted to the DNA by the UV light. The low pressure mercury vapor light emits most of its energy centered about 254 nm (1 x 10^{-5} in). This allows a high efficiency because the germicidal action curve for most organisms is centered at about 260 nm (1.02 x 10^{-5} in). Since this high energy wavelength is readily absorbed by DNA, RNA, protein, and enzymes a single photon strike affects most organisms.

Most common pathogenic microorganisms have been tested for their sensitivity to UV light, and the results have been published in various publications. *E. coli* will be reduced to a 0.0001 survival ratio, if treated in most commercially available units. *Giardia lamblia* has a survival ratio of 0.1 with twice the treatment available in most commercial units.

The sensitivity of organisms is determined by bioassay methods. Care should be exercised when using the data, because some results are achieved using various wavelengths for inactivation. Data using wavelengths other than 254 nm (1 x 10^{-5} in) may be used with corrections allowed for effectiveness.

DOSAGE

Photophysical damage is time dependent with organisms dying in a constant fraction with increasing increments of time. This is expressed as the survival ratio. The survival ratio when plotted on semilog paper is a straight line. The implications of this effect are reflected in the sizing of the unit and will be discussed later. Temperature effect on this process is negligible. The main factor in sizing is UV dosage. The recommended dosage is at least 16,000 µW-s/cm^2 (103,200 µW-s/sq in) based upon an HEW 1966 policy statement. Presently, most manufacturers treat with a dosage of 30 to 35,000 µW-s/cm^2 (190 to 225,800 µW-s/psi). These dosage rates are capable of a four-log reduction, which means that incoming water would be effectively 99.99 percent treated. Applying these numbers to the effluent of a well designed and well run sewage treatment plant would result in the treated water meeting the *E. coli* standard for drinking water.

UNIT DESIGN

Application of the available technology is quite simple and easily quantified. Most commercial units are designed to operate when supplied with water that meets a known specification. As with all drinking water supplies, the water supply must meet EPA nonbiologic standards.

Additionally, the turbidity should be less than 10 NTU. The turbidity, while not directly measuring transmission in the wavelengths under examination, is indicative of transmission. The two most commonly found UV absorbers are iron and tanin.

Most units presently available are similar in construction as far as basics. They consist of a cylindrical reaction chamber with the bulb mounted along the center axis allowing water to flow parallel to the bulb. The bulb is separated from the water by a quartz sleeve. To maintain proper treatment levels, some type of flow control device is necessary. Beyond these basics some type of monitoring and/or shut-off mechanism is required. The minimum protection would be a visible light metering device to cause an alarm or effect a shut-down in the failure mode. Utilization of the visible spectrum light for this purpose is possible because the low pressure mercury discharge tubes emit some of their energy in the visible range. Obviously, this device will not protect against all modes of failure but is sufficient for most applications. The next level of protection may be achieved by using a light sensing device that measures only light in the UV sterilization range, and has absolutely no response in any other range. This more sophisticated unit should be provided with appropriate time delays and may be slaved to any alarm or shut-down device.

APPLICATION

Most waters that do not meet the specifications can be preconditioned to acceptable levels. However, the minimum preconditioning required is a 5-µm (0.0002-in) prefilter. This is necessary to insure most particles are removed that could allow the organism's penumbral or umbral shading.

Start-up and maintenance of these units is quite simple. After installation and any breach of the system, the downstream piping should be sterilized. Normally the unit requires annual servicing, which consists of cleaning the parts and replacing the bulb.

The NSF presently has no standards concerning UV water purification. Under their proposed Standard 55, UV equipment will be accepted for the NSF seal. The proposed standard is in its first testing stages and should be established in a relatively short time.

PRECOAT CARBON FILTERS AS BARRIERS TO INCIDENTAL MICROBIAL CONTAMINATION

P. Regunathan, W. H. Beauman, and D. J. Jarog
Everpure, Inc.
Westmont, IL 60559

The microbiological characteristics of point-of-use (POU) filters have been a source of concern in the minds of many in the regulatory, utility, and academic fields. These concerns have centered around the possible growth of bacteria in various POU devices, particularly carbon filters, and the probable effect of such growths on the health of the consuming public. It is the objective of this article to investigate the role of a POU precoat carbon filter as a final barrier to microbial contamination that may otherwise inadvertently reach the consumer. A schematic view of such a possible role in the overall route of water from source to use is shown in Figure 1, giving the various acknowledged barriers that are traditionally depended upon to provide safe water to the consumer. In this article, a POU fine-filtering precoat carbon filter has been specifically examined as an added or final barrier to different types of microbial contaminants.

POU precoat filters belong to a family of POU fine-filtering devices that includes, besides precoat filters, carbon block filters made from powdered carbon, ceramic filters manufactured with controlled fine pores, pleated membrane filters with absolute ratings, and reverse osmosis membrane systems. All these devices are usually rated as capable of filtering down to 1 µm (0.00004 in) or less.

Figure 2 shows a schematic view of a precoat carbon filter where the water enters from the top through an inlet tube and rubber check valve into the bottom of the filter, making the dry powdered media there into a slurry, which is then evenly deposited as a precoat layer on the surface of a folded septum envelope. During operation, water intermittently enters the unit, filters through the precoat cake and the septum fabrics, travels through the drainage grids that are inside the septum envelope, and reaches the outlet of the filter. Precoat filter media in precoat carbon filters include mostly powdered activated carbon. Table 1 shows the significance of a precoat carbon filter as a barrier for different types of contaminants. While there are some specific barrier benefits for organic, inorganic, and particulate contaminants, this article focuses on microbial contaminants, which can be properly divided into bacterial pathogens (coliforms as indicators), viruses, protozoan cysts or surrogates, and heterotrophic bacteria. In addition, the use of silver and, in some tests copper also, has been studied as bacteriostats in precoat carbon filters.

COLIFORM REMOVAL

A test was set up to compare the abilities of filters with or without silver or copper in removing and inhibiting coliforms. Two standard precoat carbon filters, two containing copper powder added to the filter media, and two containing silver plated onto a powder and added to the media, were plumbed into a test module. The influent water was softened, dechlorinated, and fine-filtered Westmont tap water to which a nutrient-deprived suspension of *Enterobacter aerogenes* (ATCC 15038) had been added by a pump just before the module. The choices of organism, water pretreatment, and other precautions in culture preparation and transfer were designed to promote maximum coliform survival.

All six units were operated at a 0.06-l/s (1.0-gpm) flow rate, with a 30 seconds ON-30 seconds OFF cycle for eight hours per day, for several days to simulate the condition of used filters being subjected to accidental contamination. After 3,785 l (1,000 gal) had passed through each filter, samples were regularly collected just before and after overnight and weekend quiescent periods. All the data collected along with detailed procedures have been reported elsewhere (1). Results are shown graphically in Figures 3 through 6.

Influent coliform levels increased from about 10,000/100 ml (2,960/oz) to more than 100,000/100 ml (29,600/oz) due to improvements in technique. Effluent levels in standard filters (Figure 3) were approximately 99 percent lower. Both influent and effluent levels generally decreased during overnight and weekend quiescent periods. Filters with copper (Figure 4) allowed higher running coliform levels in the effluents than standard filters, probably due to disruption of the precoat cake by heavy metallic copper powder, but copper reduced these levels

Figure 1. Barriers to contamination.

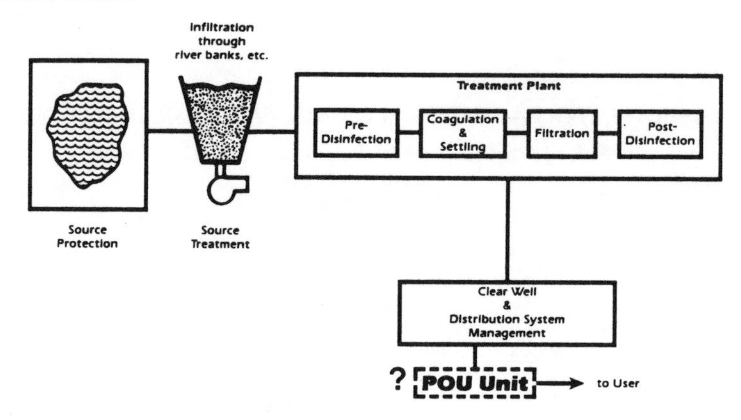

Figure 2. Precoat filter design.

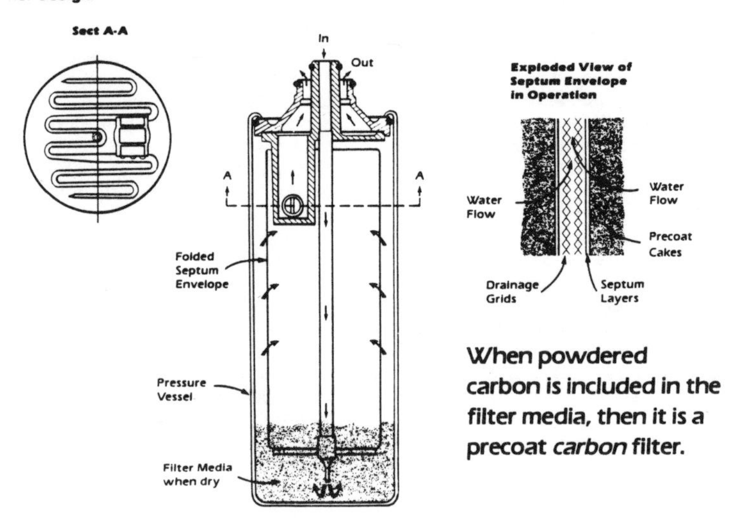

significantly during quiescent periods. Silver-containing filters (Figure 5) generally reduced the coliform levels similar to standard filters, but further exhibited significant inhibitory effect during quiescent periods. Averaged levels (Figure 6), show the relative capabilities of these filters with filters containing silver showing a further 1/2-to-1 log reduction compared to the other filters.

VIRUS REMOVAL

Standard precoat carbon filters without any bacteriostats have been recently evaluated for their ability to remove viruses using a protocol developed by a U.S. EPA-Army task force (2). Virus reduction requirements (along with bacterial and cyst reductions) for a unit to achieve purifier status have been determined by the task force to be a minimum of four-log reduction of a mixture of 10^7 plaque forming units (PFU) of poliovirus 1 and 10^7 PFU of rotavirus Wa or SA-11 per liter (0.26 x 10^7 PFU of poliovirus 1 and 0.26 x 10^7 PFU of rotavirus Wa or SA-11 per gallon).

This test on the filters was conducted at the Department of Microbiology and Immunology of the University of Arizona, Tuscon (3). Three filters were tested as per protocol (2), plumbed into a test rig served by large tanks, a pump, bladder tank, solenoid valves and timer set to operate on a cycle of three

Table 1. Significance of Contaminant Reduction

Type of Contaminant	Significance of Precoat Carbon Filter As a Barrier
Organics	
Taste and Odor-Causing Organics	Significant, but may not be health related
Volatile Organics	Not high enough capacity, low amount of carbon
Other Health-Related Trace Organics	May be significant
Inorganics	Generally no removal, unless heavy metal in precipitated form
Particulates	
Turbidity, etc.	Significant as a barrier
Microbiological	
Bacterial Pathogens or indicators	May be significant enough barrier
Viruses	May be significant enough barrier
Protozoan Cysts	May be significant enough barrier
Heterotrophic Plate Count	?

minutes ON and 27 minutes OFF for eight hours per day. Influent (control) and all three effluents were sampled seven times as per the protocol and were assayed for PFU. Complete details of procedure and results have been reported elsewhere (1,3). Table 2 shows the results of this test.

Influent levels shown in the table indicate that the high challenge levels of 10^7 PFU/l (0.26 x 10^7 PFU/gal) were not achieved in this effort. In spite of this flaw, the precoat carbon filters appear to have reduced the virus concentrations of 10^4 to 10^6 PFU/l (0.26 x 10^4 to 0.26 x 10^6 PFU/gal) in the influent to a range of <10 to 10^4/l (<2.6 to 0.26 x 10^4/gal). This reduction of approximately 99 percent is significant, even though these filters cannot be said to have achieved the four-log reduction required for purifier status. Further efforts are underway to better define the capabilities in this area.

PROTOZOAN CYST SURROGATE REMOVAL

Protozoan cysts, being inert and having no capacity for movement or reproduction, can be treated as particles that can be removed by filters at the point-of-use. There is evidence to support the use of surrogate particles in lieu of live cysts in filtration tests.

National Sanitation Foundation (NSF) Standard 53 (4) has detailed procedures for Cyst and Turbidity Reduction, and requires a filter to reduce particles in the range of 4 to 6 µm (0.00016 to 0.00024 in) by at least 99.9 percent throughout the life of the filters. These procedures were followed to test standard precoat carbon filters without any bacteriostats for their ability to remove such surrogates.

Two filters were installed in a test rig with a system of solenoid valves and a timer set to operate the units for 1.5 minutes ON and 13.5 minutes OFF for 16 hours per day, feeding Westmont tap water at 60 psig fortified with fine test dust to 20 to 30 NTU. Samples were taken at start-up and at morning start-up after the overnight quiescent periods when the initial flow rate had been reduced by filter plugging to 75, 50, and 25 percent. Samples were analyzed with a particle counter with one channel set to record 4- to 6-µm (0.00016- to 0.00024-in) particles.

Results in Table 3 show the filters to be effective in removing the 4- to 6-µm (0.00016- to 0.00024-in) particles. The filters improve in efficacy as they become plugged up with fine dust as would be expected of precoat filters.

The size of 4 to 6 µm (0.00016 to 0.00024 in) was chosen by NSF with *amoebic* and *giardial* cysts, which are two to three times larger, in mind. There has been some concern about an inadequate factor of safety in relation to *Cryptosporidium* cysts, which are also 4 to 6 µm (0.00016 to 0.00024 in) (5). New data have been gathered recently in the NSF laboratories using a submicron detector. The procedure used in this test is the Particulate Reduction test described in NSF Standard 42 (6), which is similar to the earlier test. Differences include the use of 10 minutes ON and 10 minutes OFF cycles, collection of samples at the beginning and at morning start-up after the buildup of pressure drop to 277 kPa (40 psi) across the filters, and the analysis with submicron detector of several size ranges.

Results in Table 4 show the abilities of filters to remove 0.5 µm (2 x 10^{-5} in) particles by more than 99.9 percent, indicating the ability of these filters to remove all cysts, including *Cryptosporidium* cysts.

EFFECT ON HETEROTROPHIC BACTERIA

Scientific studies have addressed concerns regarding heterotrophic bacterial growth in POU devices, notably two funded by U.S. EPA. One studied and quantified the growth in POU products (7,8), and the other studied the epidemiologic correlation of such growth with illness (9). Conclusions from these as well as other studies can be summarized to say that average increases in bacterial populations in the effluents of POU units compared to influent levels were around one order of magnitude, but exposure to such higher densities was not statistically correlated with any increase in acute symptoms, either gastrointestinal or dermatologic, compared to exposure to unfiltered water.

These studies and results, however, have not abated the concerns expressed by many (10-12). In this study, further tests have been conducted to provide additional information on the growth of heterotrophic plate count (HPC) organisms in precoat carbon filters

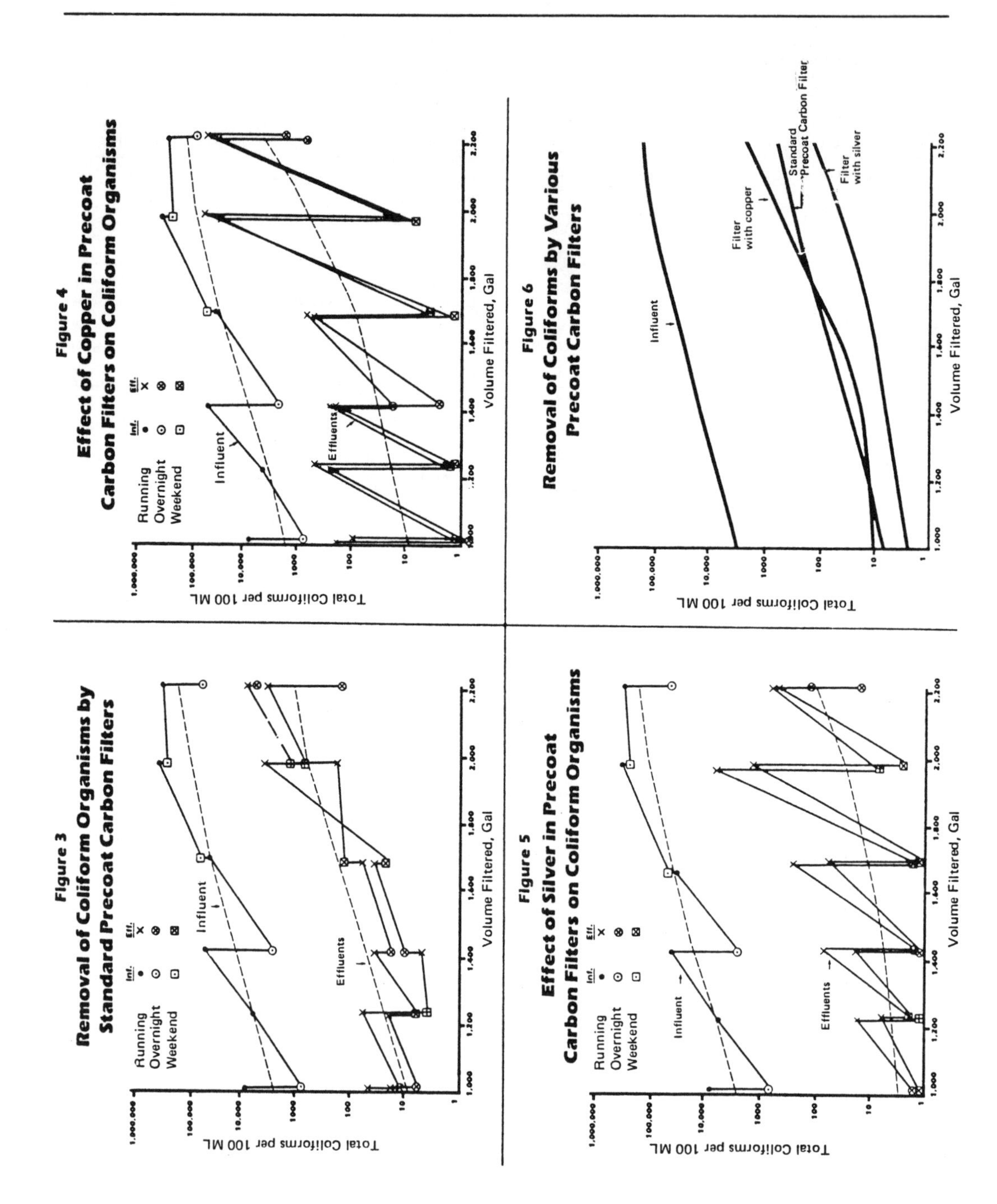

Figure 3
Removal of Coliform Organisms by Standard Precoat Carbon Filters

Figure 4
Effect of Copper in Precoat Carbon Filters on Coliform Organisms

Figure 5
Effect of Silver in Precoat Carbon Filters on Coliform Organisms

Figure 6
Removal of Coliforms by Various Precoat Carbon Filters

Table 2. Virus Removal Tests

Time (days)	Plaque-Forming Units/Liter* Influent	Unit #1	Unit #2	Unit #3	Average Reduction (percent)
1	717,000	4,200	9,000	1,400	99.32
3	230,000	2,100	6,100	3,300	98.33
6	203,000	330	330	< 10	99.89
7	257,000	670	< 10	< 10	99.91
8	60,000	< 10	< 10	< 10	> 99.98
10	53,300	< 10	< 10	< 10	> 99.92
10 (stagnant)	13,000	< 10	< 10	< 10	> 99.92

* Average of triplicates.

Table 3. NSF Cyst Reduction Test

Flow Reduction (percent)	No. of 4-6 μm Particles/ml* Influent	Unit#1	Unit #2	Average Reduction (percent)** Unit #1	Unit #2
0	76,492	76	60	99.90	99.92
25	158,140	128	92	99.92	99.94
50	348,223	35	40	99.99	99.99
75	212,783	20	19	99.99	99.99

* Average of triplicates.
** Minimum required for acceptance by NSF: 99.9 percent.

Table 4. NSF Filtration Efficiency Test

	Particle Size (μm)	Influent Counts*	Unit #1 Counts*	Unit #1 % Red.	Unit #2 Counts*	Unit #2 % Red.
Startup	0.5-1	282,165	117	99.96	245	99.91
	1-5	32,595	23	99.93	26	99.92
	5-15	395	1	NS	1	NS
	15-30	5	0	NS	0	NS
After 40 psig Δp	0.5-1	357,930	21	99.99	261	99.92
	1-5	58,640	2	99.99	7	99.99
	5-15	625	0	NS	0	NS
	15-30	10	0	NS	0	NS

* Average of triplicates.
NS - Influent challenge insufficient for significance.

and the effect of bacteriostats, specifically silver, on such growths.

It is important to point out the differences in the presently used methodologies used in HPC procedures. In earlier Standard Methods (13), plate count organisms were termed Standard Plate Count (SPC) organisms, and the procedures required the use of plate count agar with an incubation temperature of 35°C (95°F) and incubation time of two days. Many of the earlier important studies used this methodology. The most recent edition of Standard Methods (14) leaves the choice of media, incubation temperature, and time to the researcher, but requires the conditions to be spelled out along with the results. The more commonly utilized conditions appear to be R2A medium, incubation temperature of 28°C (82°F), and incubation time of seven days. The number and type of organisms enumerated under these conditions appear to be significantly higher and different from those obtained by SPC procedures. This factor was investigated in this study also.

In conjunction with the coliform removal studies discussed earlier, SPC data also were collected in the same samples that were analyzed for coliforms. Figure 7 shows all the results of the analyses indicating no concrete conclusions, except that generally the softened, carbon filtered influent waters had higher counts than the filter effluents and that silver or copper did not have any measurable inhibitory effect on these bacterial levels.

Figure 7. HPC organism in various precoat carbon filter effluents.

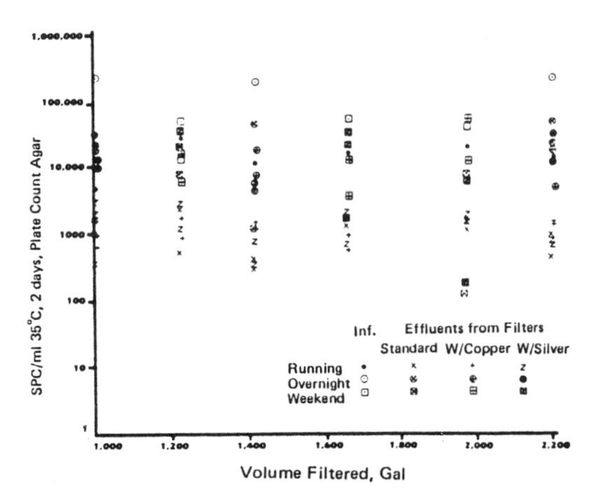

A more recent test was conducted using the Bacteriostatic Test Procedures protocol in NSF Standard 42 (6). Complete details of test procedures can be found elsewhere (1,6). Two standard precoat carbon filters and two similar filters with silvered powder added to the filter media were simultaneously tested using three different types of waters. Solenoid valves and timers were set to operate on a cycle of three minutes ON and 27 minutes OFF for 16 hours per day. The test was continued for 6670 l (1,762 gal) filtered through each unit at an operating pressure of 415 kPa (60 lb/psi) as required in the test procedures. The results are shown in Table 5. These SPC data

Table 5. NSF Bacteriostatic Test Data

Water Filtered (gal)	Type of Water Quality*	Duration of Quiescence Just Priot to Sampling-Hour	Heterotrophic Plate Counts/ml**				
				Filters w/o Silver		Filters w/Silver***	
			Influent	Unit #1	Unit #2	Unit #1	Unit #2
0	Regular	0	12	-	-	-	-
0	Regular	60	35,000	-	-	-	-
1	Regular	0	42	2	< 1	1	< 1
22	Low	8	9,000	600	1	1	< 1
73	High	8	4,400	530	< 1	1	< 1
319	Regular	8	1,900	1,300	6	5	1
740	Regular	8	3,900	97	11	1	< 1
980	Regular	8	240	36	1	1	< 1
1,220	Regular	0	11	19	< 1	1	< 1
1,220	Regular	60	23,000	3,500	3	15	1
1,460	Regular	8	135	46	< 1	1	< 1
1,722	High	8	11,000	69	< 1	1	< 1
1,762	Low	8	3,200	1,400	430	330	4

* Regular - 200-600 ppm TDS, 7.2 ± 0.5 pH.
Low - 25-100 ppm TDS, 6.2 ± 0.5 pH.
High - >800 ppm TDS, 9.5 ± 0.5 pH.
** Duplicate analyses, 35°C, 48 hr.
*** Effluent silver; 1-3 ppb.

indicated that both the units without silver, as well as those with silver were bacteriostatic filters, according to NSF Standard (6) requirements. Silver, however, does seem to provide some extra assurance and capability in providing fairly low counts in the effluents of both units, while one of the two units without silver had significantly higher counts than its duplicate unit.

During July 1987 a comprehensive test program was initiated to study the growth of SPC and HPC organisms in standard precoat carbon filters with or without silver as a bacteriostat, and in granular carbon bed units. The following were the pertinent factors controlled or used in the tests:

- Three of each type of unit (standard, w/silver, granular) with two influent locations.
- 0.03 l/s (0.5 gpm) through each filter and each influent port.
- Westmont municipal water as supplied.
  Total chlorine content in running samples = 1.0 to 1.5 mg/l (1.0 to 1.5 ppm)
  Free chlorine content in running samples = trace levels.
- Three minutes ON, 27 minutes OFF, eight hours per day only during working days.
- Sampling
  - Monday AM at start-up
  - Wednesday PM during running
  - Thursday AM at start-up
  - Friday PM during running
- 1 l (0.26 gal) sample ($\frac{1}{2}$ minute run) collected. Bacteriological sample from this into sterile bottle with Chambers' neutralizing solution. (Influent samples always taken before filter effluents.)
- Only R2A medium used in the beginning. Started with pour plates, switched to spread plates. Started with 25°C (77°F) for five days, changed to 28°C (82°F), seven days.
- Tests started 7/17/87.
- Plate count Agar, incubation at 35°C (95°F) for two days routinely used from 9/2/87 for SPC measurements.

HPC data collected using R2A media for the influent samples and the three sets of effluent samples are graphically shown in Figures 8 through 11. Arithmetic means of all these separate sets of data have been calculated and are shown in Table 6. Comparison of the data in these figures and table would indicate that the use of silver in precoat carbon filters results in maintaining the organism levels at or below the influent levels for the different types of samples, i.e., running, overnight, and weekend samples. Running and overnight quiescent samples from standard precoat carbon filters without silver appear to have 1 to 1-1/2 orders of magnitude higher HPC counts than influent samples or the effluent samples from units with silver. Granular carbon filter effluents yielded generally 1-1/2 orders of magnitude higher counts than influent waters or precoat units with silver, when running or overnight samples were

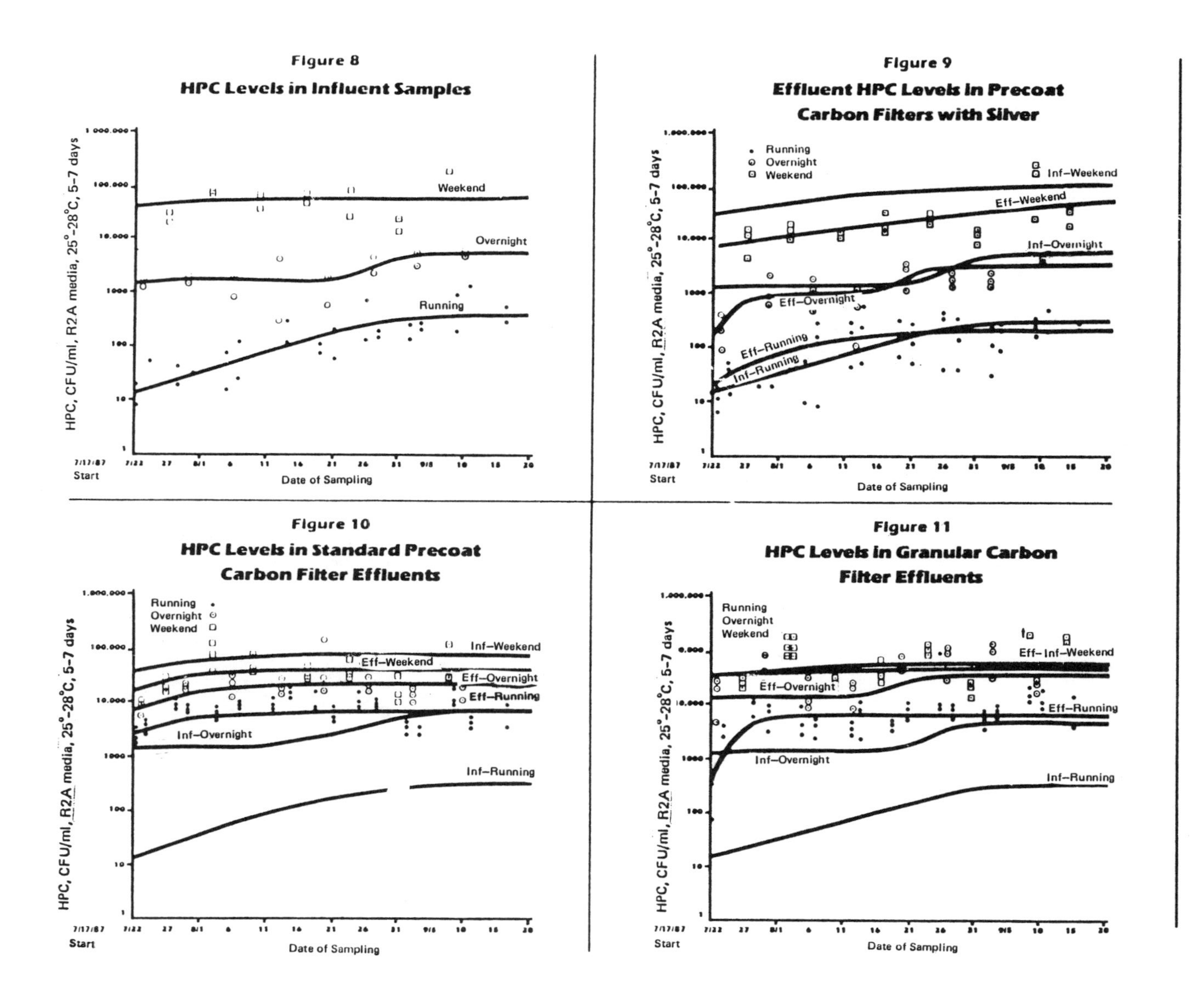
Figure 8
HPC Levels in Influent Samples
HPC, CFU/ml, R2A media, 25°-28°C, 5-7 days
Weekend
Overnight
Running
7/17/87
Start
Date of Sampling
Figure 9
Effluent HPC Levels in Precoat Carbon Filters with Silver
Running
Overnight
Weekend
Inf-Weekend
Eff-Weekend
Inf-Overnight
Eff-Overnight
Eff-Running
Inf-Running
Figure 10
HPC Levels in Standard Precoat Carbon Filter Effluents
Inf-Weekend
Eff-Weekend
Eff-Overnight
Eff-Running
Inf-Overnight
Inf-Running
Figure 11
HPC Levels in Granular Carbon Filter Effluents
Eff-Inf-Weekend
Eff-Overnight
Eff-Running
Inf-Overnight
Inf-Running

compared. Weekend quiescent samples from influent ports and granular carbon filters were in the same order of magnitude, while precoat units with or without silver generally yielded 50 percent lower numbers.

Table 6. Comparison of Bacteriological Data, Arithmetic Mean HPC*, CFU/ml

	Influent	Silver	Standard	Granular
Running	194	260	7,040	8,730
Overnight	2,510	1,870	20,900	44,700
Weekend	75,700	27,600	37,300	87,200

* R2A media, 25-28°C, 5-7 days.

Table 7. Comparison of Bacteriological Data, Arithmetic Mean SPC*, CFU/ml

	Influent	Silver	Standard	Granular
Running	33	47	23	1,050
Overnight	444	332	515	4,180
Weekend	26,700	7,350	7,240	22,600

* SPC: Pour Plates, 35°C, 2 days.

SPC data collected using plate count agar (35°C [95°F], two days) for similar units are graphically shown in Figures 12 through 15. Arithmetic mean values for the separate sets of data have been calculated and are shown in Table 7. Examination of these values shows different conclusions from those reached using HPC data. Unlike earlier tests, these data show that precoat filter units with and without silver yield organism counts equal to or lower than influents when running water, overnight quiescent, or weekend quiescent samples are compared between themselves. Only granular carbon bed units yield 1 to 1-1/2 orders of magnitude higher counts in running and overnight quiescent samples.

Table 8 shows the direct comparison of HPC and SPC data from all the samples for which both analyses were performed. This comparison confirms earlier stated observations and conclusions. Further, it shows that R2A media at 28°C (82°F) for seven days yields 1/2 to 1-1/2 orders of magnitude higher counts than those obtained using plate count agar at 35°C (95°F) for two days. The regulatory efforts to control these organisms at a particular level need to consider these huge differences in values obtained from identical samples.

These comparisons indicate that silver when used in precoat carbon filters has a selective effect on organisms. Earlier coliform removal tests showed that silver had a measurable effect on *Enterobacter aerogenes*. Silver also seems to have a significant effect on those organisms that grow in R2A media at 28°C (82°F), while it has no measurable effect on those organisms that grow in plate count agar at 35°C (95°F). Efforts to identify these selective effects were not successful, because many of the colonies on the plates could not be identified. The few identifiable organisms shown in Table 9 did not yield enough useful information. Further efforts are needed in this area of activity.

CONCLUSIONS

On the basis of this study, the following conclusions can be reached:

- Data presented show significant and consistent reductions by these precoat filters of coliforms (~99 percent), enteric viruses (~99 percent), and protozoan cyst/surrogates (>99.9 percent).

- Silver in precoat carbon filters lowers coliform levels at least 1 log more than the standard precoat filters. Silver and copper act slowly to reduce coliform levels in filters during non-use periods.

- The decision to use R2A agar at 25 to 28°C (77 to 82°F) for seven day incubation procedures instead of SPC agar pour plates at 35°C (95°F) for two days is not trivial. Not only does the new HPC procedure yield 1/2 to 1-1/2 logs higher counts, it also favors a different population and thus can lead to different overall conclusions.

- Silver appears to effectively control the filter effluent HPC levels that are found growing in R2A media incubated at 25 to 28°C (77 to 82°F). Proper comparisons of all data indicated silver as bacteriostatic in precoat carbon filters.

- Precoat filters with or without silver appear to act as barriers to incidental microbial contamination because they do confer some measurable level of protection against incidental microbial contamination of potable water supplies that are normally safe.

ACKNOWLEDGEMENT

The authors thank Charles Ferrara for producing the illustrative and graphic figures, Robert Gonzalez for collecting and assembling data, and Pam Menefee for producing the manuscript.

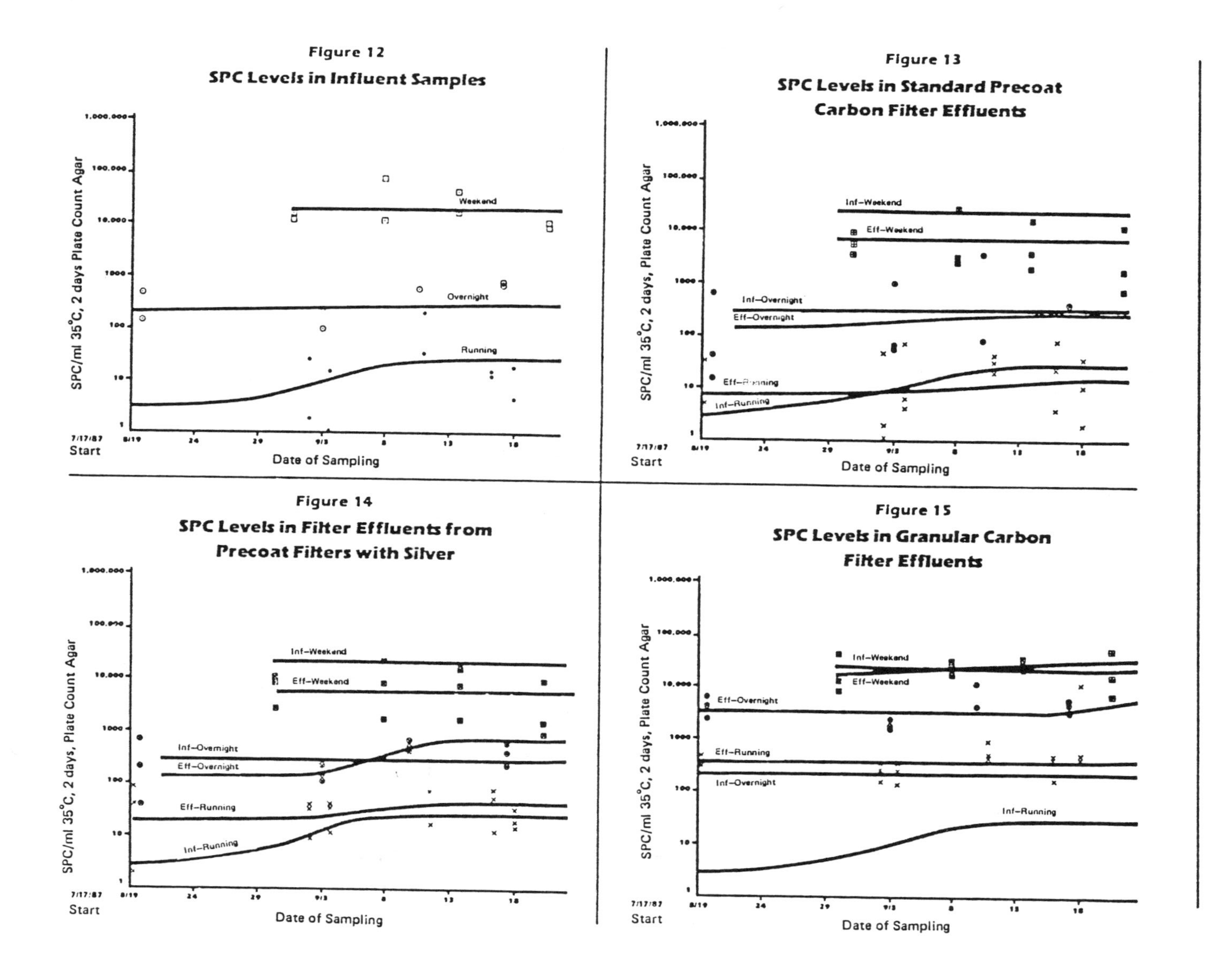

Figure 12
SPC Levels in Influent Samples

Figure 13
SPC Levels in Standard Precoat Carbon Filter Effluents

Figure 14
SPC Levels in Filter Effluents from Precoat Filters with Silver

Figure 15
SPC Levels in Granular Carbon Filter Effluents

Table 8. Direct Comparison of SPC vs. R2A, Arithmetic Mean, CFU/ml

	SPC: Pour Plates, 35°C, 2 days				HPC: R2A Spread Plates, 25-28°C, 5-7 days			
	Influent	Silver	Standard	Granular	Influent	Silver	Standard	Granular
Running	33	47	23	1,050	331	497	4,880	8,050
Overnight	444	332	515	4,180	3,410	3,130	29,200	60,200
Weekend	26,700	7,350	7,240	22,600	108,000	46,700	34,900	126,000

Table 9. Documented Organisms

	Influent	Standard	Silver	Granular
Pseudomonas				
acidovrans	X	X		
luteola	X	X	X	X
paucimobilis	X	X	X	X
vesicularis	X	X	X	X
stutzeri	X	X		X
cepacia	X			
fluorescens		X		
mallei				X
Flavobacterium spp.	X	X	X	X
Agrobacterium radiobacter	X	X	X	
Achromobacter spp.		X		

REFERENCES

1. Regunathan, P. and Beauman, W. H. Microbiological characteristics of point-of-use precoat carbon filters. JAWWA, 79:10:67, October 1987.

2. Gerba, C. P. and Thurman, R. Towards developing standard procedures for testing microbiological water purifiers. Proceedings of the Third Conference on Progress in Chemical Disinfection. Binghamton, NY, April 1986.

3. Gerba, C. P. and Kutz, S. M. Evaluation of Everpure 4C cartridge filters for virus removal. Unpublished report, University of Arizona, Tucson, May 1987.

4. National Sanitation Foundation. Standard 53: drinking water treatment units -- health effects. Ann Arbor, MI, Rev. June 1982.

5. Culotta, N. J. Personal communication. National Sanitation Foundation, Ann Arbor, MI, March 27 1987.

6. National Sanitation Foundation. Standard 42: drinking water treatment units -- aesthetic effects. Ann Arbor, MI, Rev. June 1982.

7. Criteria and Standards Division, Office of Drinking Water, U.S. EPA. Fact sheet/update, home water treatment units contract. July 1980.

8. Bell, F. A., Perry, D. C., Smith, J. K. and Lynch, S. C. Studies on home treatment systems. JAWWA 76:2, 126-130, February 1984.

9. Calderon, R. L. An epidemiological study on the bacteria colonizing granular activated carbon point-of-use filters. In Press: Proceedings of the Water Quality Association Annual Conventions. Dallas, TX, March 1987.

10. Geldreich, E. E. et al. Bacterial colonization of point-of-use water treatment devices. JAWWA 77:2, 72-80, February 1985.

11. Reasoner, D. J. et al. Microbiological characteristics of third-faucet point-of-use devices. JAWWA, 79:10:60, October 1987.

12. Fed. Reg. 50:219, November 13, 1985.

13. Standard methods for the examination of water and wastewater. APHA, AWWA, and WPCF. Washington, D.C. (15th ed., 1980).

14. Standard methods for the examination of water and wastewater. APHA, AWWA, and WPCF. Washington, D.C. (16th ed., 1985).

MICROBIOLOGICAL STUDIES OF GRANULAR ACTIVATED CARBON POINT-OF-USE SYSTEMS

Donald J. Reasoner
Drinking Water Research Division
Water Engineering Research Laboratory
U.S. Environmental Protection Agency
Cincinnati, OH 45268

INTRODUCTION

Activated carbon, powdered (PAC) or granular (GAC), has been used in water treatment for many years, primarily to remove tastes and odors. During the past 15 to 20 years, the availability and sales of small PAC and GAC filter units, or point-of-use (POU) treatment units, have increased steadily. Much of the increased demand for these units has resulted from consumer concerns over the quality of the water supplied by local water utilities. These concerns are stimulated by national and local news reports of organic and inorganic chemicals in drinking water, health risks due to long-term ingestion of potentially carcinogenic compounds in drinking water, taste and odor problems in the local water supply, and other problems of treatment and/or distribution. Other factors contributing to the use of POU devices include lack of a centrally treated water supply, contaminated ground water supplies, and ground water supplies containing naturally high concentrations of iron, sulfur, nitrates, or fluorides. An aspect of GAC POU treatment devices that has been of concern to the Drinking Water Research Division, Water Engineering Research Laboratory, U.S. EPA, Cincinnati is the long-term microbiological quality of the product water from such devices.

The first phase of studies on the microbiological characteristics of carbon POU filters was begun in our laboratories in 1977. This phase examined four carbon POU filters for variations in bacterial counts, heterotrophic plate count (HPC), total organic carbon (TOC), and chloroform ($CHCl_3$) levels. In addition, one filter was installed on a low-flow drinking water fountain, and bacteriological quality and THM content of the product water were monitored weekly. HPC levels in the effluents from the carbon filters were always higher than the bacterial levels of the influent tap water. One of the test filters had consistently higher HPC levels than the other three test filters, and morning samples generally had higher HPC levels than did those in the afternoon. Free-chlorine removal for the test carbon filters ranged from about 53 to 65 percent in one run, and 77 to 97 percent in a second run. Total organic carbon removal appeared to be minimal because influent TOC levels were always less than 2 mg/l (2 ppm) and effluent levels were within 0.1 to 0.3 mg/l (0.1 to 0.3 ppm) of the influent levels one day after filter installation. Two to four weeks later, effluent levels were essentially the same. Chloroform removal effectiveness was directly related to the amount of carbon in the filter. The initial percentage removal of $CHCl_3$ ranged from 100 percent down to about 55 percent, and decreased with time over a period of 20 weeks. By the end of the 20-week test period, effluent $CHCl_3$ concentration for three of the test filters exceeded influent $CHCl_3$ concentration, while the fourth filter unit was still removing about 20 percent of the influent $CHCl_3$.

The second phase examined four additional GAC POU filters designed to be installed as stationary filters under a kitchen sink on the cold water line to a common mixing faucet. This study examined the HPC of the product water from the POU filters and the potential colonization of the test filters when challenged by pure culture suspensions of opportunistic and frank bacterial pathogens that might contaminate a potable water supply. This phase of the POU study was reported by Geldreich et al. (1). Variations in HPC levels in the product water between morning and afternoon samples from the test filters were similar to those found during the first phase study. Additionally, it was shown that stagnation periods from several hours to several weeks resulted in significantly increased HPC levels in the test filters. The HPC levels in the product water could be reduced after a stagnation period by simply flushing the units thoroughly at full-flow for two to three minutes before using the water.

The potential for colonization of the test filter units was examined by using pure culture suspensions of *Serratia marcescens*, *Pseudomonas aeruginosa*,

Enterobacter cloacae, *Enterobacter aerogenes*, *Escherichia coli*, *Citrobacter freundii*, and *Salmonella typhimurium*. *E. coli*, *S. typhimurium*, and *E. cloacae* did not colonize the filters and were not detected in the product water after the initial sample. *S. marcescens*, *P. aeruginosa*, *E. aerogenes*, and *C. freundii* persisted and were found in the product water from some or all of the test filters for periods of time ranging from five days for *C. freundii* to 156 days for *S. marcescens* and *P. aeruginosa*.

In addition to following the HPC levels of the POU filter units and challenging them with the organisms above, new filter cartridges were installed, disinfected, and the product water monitored for 12 months for HPC bacteria. Bacterial cultures were periodically isolated, purified, and identified. The organisms found during this 12-month period included *C. freundii*, *E. aerogenes*, *E. cloacae*, *K. pneumoniae*, *Alcaligenes spp.*, *Pseudomonas cepacia*, *P. fluorescens*, *P. maltophila*, *S. marcescens*, and *S. rubidaea*. Not all of these organisms were isolated from each of the filter units at the same sampling time, and no individual unit yielded isolates of all of the organisms. Only *E. cloacae*, *Alcaligenes spp.*, and *P. maltophila* were found in all the units at some time during the 12-month period, and none of the isolates was found to be continuously present. Since the POU test set-up had been disinfected prior to the beginning of this 12-month study, the organisms isolated must have been present in the treated distribution water influent to the test system, and were able to survive and multiply to some extent in the POU filter cartidges.

The third phase of the GAC point-of-use study, currently nearing completion, was designed to examine GAC POU treatment units intended to be installed as third-faucet units, thus treating only water for drinking and cooking, not all of the water going to the main kitchen faucet. Seven test units were selected for this phase. The test set-up configuration from the previous two study phases was retained, but the plumbing was modified to accept the additional three test units. Partial results from this phase were reported earlier as technical conference presentations and appeared in the October issue of the Journal of the American Water Works Association. HPC results from this phase of the POU study showed several types of variation. POU product water HPC levels varied depending on the time of sample collection (morning versus afternoon). Generally, the afternoon samples contained fewer HPC bacteria than did the morning samples, reflecting the effect of flushing on wash-out of bacteria during the daytime simulated use periods. The magnitude of the difference beween morning and afternoon HPC levels varied among the test units. Examination of the variation in the monthly mean HPC indicated that there was no single pattern of HPC results for all the test units other than the changes that occurred with flushing between the morning and afternoon samples. Some of the variation in HPC monthly means may be attributed to the combined influence of seasonal changes in water temperature and changes in disinfectant residual. HPC levels for some test units clearly decreased as water temperature decreased from September through December. Peaks in HPC levels usually occurred in July, August, or September, corresponding to peak water temperatures.

Some HPC variation may reflect the influence of unit design, volume of GAC in the cartridge, and possibly the material used in the construction of the cartridge holder and the cartridge itself. Metal cartridge housings may contribute to more rapid equilibration of the water and GAC within those units to ambient room temperature during nonflow periods. Any increase in water and GAC cartridge temperature would result in increased growth of bacteria on the GAC, resulting in higher HPC levels in the product water.

The bacterial flora (HPC) of the dechlorinated tap water influent to the test POU units reflected the contribution of the GAC dechlorinating filter. This filter served to remove free chlorine from the distribution water and seeded the dechlorinated water with bacteria. Thus, reasonable worst case conditions were set up for the tests (i.e., no free chlorine residual and HPC levels greater than the treated distribution water). The mean HPC of the dechlorinated tap water was generally lower than that of the water from the best of the POU test units. POU test units that included silver as a bacteriostatic agent were found to have HPC levels as variable as the nonbacteriostatic POU test units. The bacterial flora of the bacteriostatic units appeared to be different, both in colony appearance and variety, from the bacterial flora of the nonbacteriostatic POU units. The silver served as a selective agent, allowing silver-tolerant bacterial strains to grow.

All of the POU test units modified the percentage of pigmented bacteria found in the product water, as compared to the percentage of pigmented bacteria found in the dechlorinated tapwater. Generally, the percentage of pigmented bacteria present in water from the bacteriostatic POU test units was lower than that of the nonbacteriostatic test units. The dechlorinated tap water usually contained more than 50 percent pigmented bacteria, whereas the pigmented bacterial content of the water from the POU units ranged from less than four percent to about 40 percent depending on whether the samples were taken in the morning or afternoon. Old filters (on line for several months) tended to show relatively stable HPC and pigmented bacterial levels, whereas newly installed filters had low initial HPC levels that rapidly increased during the first two to three weeks of use, and became fairly stable thereafter.

Challenges of test POU filter units with specific bacterial pathogens (*Klebsiella pneumoniae*,

Aeromonas hydrophila, and *Y. enterocolitica*) during this third phase study showed that only K. pneumoniae colonized the test filters for an extended period of time (2). *Aeromonas hydrophila* colonized the POU filters during a warm water (20°C, October 1984) experiment but not during a cooler water (12°C, February 1986) period. Recently, challenge experiments with *Legionella pneumophila* were concluded. The results of these experiments indicated that *L. pneumophila* apparently did not colonize the POU test filters. However, recovery methodology for *L. pneumophila* is not efficient and lacks sensitivity at low cell concentrations, and it cannot be stated with certainty that this organism will not colonize GAC POU treatment units. Cool or cold water temperatures and the presence of any disinfectant residual (total chlorine residual) in the water probably mitigate against colonization by this organism.

The general implications of the studies conducted in our laboratory, as well as other published studies, are that all GAC POU devices become generators of bacteria due to the large surface area exposed to the water, and due to adsorption of nutrients from the water that bacteria are able to use for growth. However, the potential for adverse human health effects from ingestion of large numbers of HPC bacteria in water appears to be low. To date, there have been no verified reports of waterborne illness resulting from consumption of contaminated water from GAC or other POU treatment devices.

Practical recommendations for users of home POU treatment devices are as follows:

- Use the POU device only on a microbiologically safe water supply, unless specifically recommended by the manufacturer for other applications as well.

- Prior to using the product water from the POU device after a prolonged quiescent period (several hours or overnight), run the water to waste for 30 seconds or longer at full flow. Longer flushing is desirable after a prolonged nonuse period such as a vacation.

- Change the filter cartridge(s) at least as frequently as recommended by the manufacturer, or preferably more often.

- Adhere to the manufacturer's maintenance recommendations and specific instructions relative to changing the filter cartridge(s).

REFERENCES

1. Geldreich, E. E., Taylor, R. H. Blannon, J. C. and Reasoner, D. J. Bacterial colonization of point-of-use water treatment devices. JAWWA. 77:72-80, 1985.

2. Reasoner, D. J., Blannon, J. C. and Geldreich, E. E. Microbiological findings with point-of-use, third faucet devices. JAWWA. 79, October 1987.

HEALTH STUDIES OF AEROBIC HETEROTROPHIC BACTERIA COLONIZING GRANULAR ACTIVATED CARBON SYSTEMS

Alfred P. Dufour
Toxicology and Microbiology Division
Health Effects Research Laboratory
U.S. Environmental Protection Agency
Cincinnati, OH 45268

Aerobic heterotrophic bacteria are ubiquitous in the aquatic environment. Since surface waters frequently serve as a source for potable water, it is not unusual to find members of this large heterogeneous group in drinking water. The types and species of heterotrophic bacteria found in drinking water have been described by various authors and, in general, the organisms are gram negative, nonspore-forming bacilli. The second column of Table 1 is a composite list of bacteria isolated from raw and treated drinking water. This list is for comparison purposes only, and does not represent all those heterotrophic bacteria that are indigenous to drinking water distribution systems. These bacteria occur in drinking water at densities as high as several hundred thousand per milliliter in some cases, and they provide a constant seed for devices used to treat potable water at its point-of-use. Granular activated carbon (GAC) filters, in this regard, are of special interest because of their increased use by the general population and because they have the capacity to adsorb bacteria from water. Once adsorbed on the GAC filter, the bacteria are able to multiply to even higher densities than observed in the influent water and they, in turn, slough off into the effluent water. The genera and types of bacteria found on GAC filters are listed in column 3 of Table 1, and it is obvious from this list that the bacteria isolated from filters reflect the distribution of bacteria observed in drinking water systems. Gram negative bacteria are predominant, just as in the water distribution systems. The density of bacteria observed in GAC filter effluents has been in the hundreds of thousands, and this frequently represents an increase of many magnitudes above the density of heterotrophic bacteria in the influent water.

The amplification of heterotrophic bacteria by GAC filters has caused some concern that these bacteria may pose a health risk to water users. The reason for this concern is the infrequent observation that some of the heterotrophic bacteria isolated from drinking water and GAC filter effluents have been associated with nosocomial or hospital associated infections and illnesses. The fourth column of Table 1 is a partial list of the types of bacteria causing infection and illness under hospital conditions, and which were thought to be due to contact with drinking water. The water source, in most cases, was some type of amplifier, which increased the densities of the organism in the water that was linked to the patient's illness or infection. The amplifiers were varied and included devices such as humidifiers (3,4), dialysis machines, disinfectant bottles, and water reservoirs for pediatric isolettes (7). The infections ranged from *septicemia* to pneumonia to peritonitis and, in some cases, have been fatal. In most, if not all, of the cases the normal body defense mechanisms of the patient had been compromised in some way. The bacterial isolates in column 4 of Table 1 were obtained from patients who suffered from diabetes (5), bronchitis (6), had just

Table 1. Bacterial Isolates Associated with Raw and Treated Potable Water, Granular Activated Carbon Filters, and Nosocomial Infections

	Source of Bacterial Isolates		
Bacterial Type	Raw/Treated Potable Water (Ref. 1,2)	Granular Activated Carbon (Ref. 2)	Nosocomial Infection (Ref. 3-8)
Acinetobacter	X	X	X
Aeromonas	X		
Alcaligenes	X	X	
Citrobacter	X	X	
Enterobacter	X	X	
Klebsiella	X	X	X
Moraxella	X	X	
Pseudomonas	X	X	X
Serratia	X		X
Flavobacterium	X	X	X
Staphyloccus	X	X	
Bacillus	X		
Achromobacter			X

undergone open-heart surgery (8), or were on a regimen of steroid therapy (4). The patients in two cases, were infants whose immunological systems had not yet matured (3,7). The linkage of nosocomial infections to drinking water has been instrumental in part in promoting the concern about heterotrophic bacteria in drinking water and GAC filters. The conclusions drawn from nosocomial infections may, however, be very misleading in that they do not address the fact that the patients were usually compromised in some manner. Infections of this type are usually caused by bacteria that are avirulent or have limited virulence, and which seize the opportunity offered by weakened defense mechanisms to inflict damage to the host. These bacteria are called opportunistic pathogens, and they seldom cause illness in healthy individuals. Although the available evidence indicating that healthy individuals are not at risk from these bacteria appears to be strong, there is no empirical data supporting this conclusion. Since the Environmental Protection Agency (EPA) may be placed in the position of recommending GAC filters as an alternative form of water treatment for removing organic chemical hazards present in drinking water, it is necessary to know whether or not that hazard is being replaced by another (i.e., heterotrophic bacteria amplified by the filter). An epidemiological study conducted by Yale University was supported by the EPA in order to determine if adverse health effects are associated with GAC filter use. The results of the Yale study are reviewed here to characterize the risks observed in healthy populations exposed to water treated with point-of-use GAC filters (9,10).

The study was conducted at a large military reservation in eastern Connecticut. Families with children were recruited for the study from the large residental population of 800 families at this naval base. A military base population was considered ideal because the study participants had easy access to cost-free medical care. This factor was critical since a laboratory workup of clinical specimens obtained from participants was a requirement for each illness or infection where an individual consulted a physician.

The health reporting aspect of the study used two approaches. The first approach used the calendar system. Each participating family in this system was given a health status calendar form on which they could fill in, on a daily basis, their health status. Symptomatology, such as vomiting, nausea, diarrhea, high temperature, skin infections, and rashes, as well as visits to a doctor were recorded. This was usually done by the mother of the family or some other responsible adult. The calendars were usually collected every two weeks so that any participants who failed to keep the calendar up-to-date could be questioned and their previous two-week health status recorded. The second approach involved instructing each participating family to go to their medical facility if they experienced gastrointestinal illness or skin infections. When a gastrointestinal illness or skin infection resulted in a clinical bacterial isolate, the filter associated with the individual from whom the isolate was obtained would be replaced with a new filter and the old GAC filter bed would be analyzed to determine if a bacterial specie could be isolated that would match the clinical isolate.

Two types of filters were used in the study. One was a faucet-type filter that attaches to the tap with a special adapter. The filter was activated by turning a valve that directed the water through the GAC bed. The second type of filter was a bypass-type that tapped directly to the cold water line and delivered the water through a separate tap attached to the sink. The source of the water serving the study population was the Groton, Connecticut city water supply. The water is obtained from the Great Brook watershed, and it is sand filtered before being sent into the distribution system.

Two bacteriological media were used to assay the water samples during the course of the study. Aerobic heterotrophs were enumerated using Standard Plate Count agar (11) and the R2A medium of Reasoner and Geldreich (12). A summary of the results of the bacterial monitoring is shown in Table 2.

It is immediately apparent that the R2A agar detected much higher densities of heterotrophic bacteria than the Standard Plate Count agar. This observation has been noted by others, and it is thought to occur because the two media detect different parts of the distribution of heterotrophic bacteria. One other interesting result of the monitoring is the high initial densities of heterotrophs observed in the water samples taken from the faucet filter housing units without carbon. The cause of these high densities is unknown. It is possible that the housing units were contaminated before installation, however, this effect did not show up in filters with carbon beds. The greatest exposure to heterotrophic bacteria occurred with the bypass filters. The exposure to heterotrophic bacteria was, on the average, about 20 times greater for bypass-filter users than for control groups exposed to heterotrophic bacteria in unfiltered tap water. The faucet-type filter effluents contained about 12 times more heterotrophs than were found in the tap water. However, the heterotrophic bacterial densities in the bypass and faucet-filter effluents exceeded that of the blank filter housing units by factors of only six and four, respectively. Thus, the exposure differences were not as great when the faucet housing only group was used as a control population, as was the case in this study. The means of the heterotroph densities of both the faucet-type and bypass-type filters were statistically significant from the mean density of heterotrophs observed in the blank housing unit effluents.

Table 2. Comparison of Bacterial Densities in Tap Water and GAC Filter Effluents on Standard Plate Count Agar and R2A Agar

Filter Type	Geometric Mean of Heterotrophic Bacteria* (Ref. 10)			
	SPC Agar		R2A Agar	
	Initial	Subsequent	Initial	Subsequent
Bypass	0.2 (62)	1049 (722)	0.4	2,042
Faucet	6.0 (10)	689 (559)	9.0	1,035
Faucet Housing (no carbon)	85.0 (43)	198 (486)	98.0	289
Tap Water	6.0 (215)	53 (1,776)	11.0	92

* Numbers in parentheses indicates number of samples analyzed.

During the course of the study only a few individuals reported to the medical faciiity for treatment. No bacterial specimens were obtained from these patients and, therefore, no relationship between clinical isolates and isolates from GAC could be established.

A summary of the data collected using the calendar questionnaire system is shown in Table 3. This system serves a two-fold function. First, it may detect excess illness in filter users that might not have been observed clinically and, second, it may detect a decrease in illness among filter users, if the filters adsorb potential nonbacterial pathogens that might occur in the source water. The data in the table is given in terms of the number of symptomatic illnesses that occurred per thousand person years of filter usage. The total person years of usage was 230 for the faucet-type filters and 181 for those that did not use filters. The data were analyzed statistically, and no significant differences in the rates of symptomatic illnesses were observed between the two user groups and the control group. Thus, even with the questionnaire data, it could not be shown that excess illness could be linked to the use of GAC filters. Conversely, there was no evidence that the illness rate was lowered in GAC filter users.

Table 3. Comparison of Symptomatic GI Illness Rates Observed in Study Groups Using GAC Point-of-Use Filters and a Control Group

Sympton	Symptomatic Illness Rate/1,000 Person-Years in Groups (Ref. 10)		
	Bypass Filter	Faucet Filter	None
Vomiting	32	36	33
Nausea	45	51	49
Diarrhea	59	76	74
Fever	58	47	62
Body Aches	46	46	55
Skin Rash	13	12	10
Infected Wound	1	3	3

The conclusion that can be drawn from the results of the Yale study is that point-of-use granular activated carbon treated water containing high densities of heterotrophic bacteria is not a risk factor for healthy populations.

The Environmental Protection Agency is supporting further research on the use of GAC filters and health effects which will be conducted by Yale University. The research reviewed here addressed exposure to high densities of heterotrophic bacteria via the ingestion route. Point-of-entry type filters add a new dimension to potential exposures since all of the water entering a home is treated. Amplified heterotrophic bacterial densities can be disseminated in aerosols from showerheads and, subsequently, carried into the body via the respiratory route. The results of the continuing study may provide some information on the etiology of respiratory illness in the United States.

REFERENCES

1. LeChevallier, M. W., Seidler, R. J. and Evans, T. M. Enumeration and characterization of standard plate count bacteria in chlorinated and raw water samples. Appl. Environ. Microbiol. 40:922, 1980.

2. Parsons, F. Microbial flora of granular activated carbon columns used in water treatment. In: Wood, P. R., Jackson, D. F., Gervers, J. A., Waddell, D. H. and Kaplan, L., Removing Potential Organic Carcinogens from Drinking Water, Vol. I. Appendix A, U.S. Environmental Protection Agency, EPA-600/2-80-130a, Cincinnati, OH, 1980.

3. Foley, J. F., Gravelle, C. R., Englehard, W. E. and Chin, T. D.·Y. *Achromobacter Septicemia* - fatalities in prematures. Amer. J. Dis. Children. 101:279, 1961.

4. Smith, P. W. and Massanari, R. M. Room humidifers as a source of *Acinetobacter* infections. J. Amer. Med. Assoc. 237:795, 1977.

5. Berkleman, R. L., Godley, J., Weber, J. A., Anderson, R. L., Lerner, A. M., Peterson, N. J. and Allen, J. R. *Pseudomonas cepacia* peritonitis associated with contamination of automatic

peritoneal dialysis machines. Ann. Int. Med. 96:456, 1982.

6. Mertz, J. J. Scharer, L. and McClement, J. H. A hospital outbreak of *Klebsiella* pneumonia from inhalation therapy with contaminated aerosol solutions. Amer. Rev. Resp. Dis. 95:454, 1967.

7. Scheidt, A., Drusin, L. M., Krauss, A. N. and Machalek, S. G. Nosocomial outbreak of resistant *Serratia* in a neonatal intensive care unit. N.Y. State J. Med. 82:1188, 1982.

8. Herman, L. G. and Fournelle. Flavobacteria: a water-borne potential pathogen. In: Proceedings of the Third International Congress on Chemotherapy, Stuttgart, Germany, July 22-27, 1964.

9. Mood, E. W. and Calderon, R. L. An epidemiological study on bacteria in point-of-use activated carbon filters. Draft Report to Health Effects Research Laboratory, Cincinnati, OH for Cooperative Agreement CR-811904, 1987.

10. Calderon, R. L. An epidemiological study on the bacteria colonizing granular activated carbon point-of-use filters. Point-of-Use. 5:1, 1987.

11. Standard methods for the examination of water and wastewater. Amer. Public Health Assoc., Washington, DC, 15th ed, 1981.

12. Reasoner, D. J. and Geldreich, E. A. A new medium for the enumeration and subculture of bacteria from potable water. Appl. Environ. Microbiol. 49:1, 1985.

ACTIVATED ALUMINA FOR POU/POE REMOVAL OF FLUORIDE AND ARSENIC

Robert L. Lake
Water Treatment Engineers
Scottsdale, AZ 85257

INTRODUCTION

For the past 14 years, we have been treating high fluoride and/or arsenic water with activated alumina (AA) in POU/POE applications. Typical systems are: 1) a unit on each drinking fountain at a school; 2) a unit for each home in a subdivision of over 300 homes; 3) a dual system for a 365-room hotel in combination with a 185-unit trailer park; 4) a single water tap for a small trailer park; and 5) all potable water in a restaurant.

Each system may require a different approach with respect to installation, monitoring, and servicing. We are currently the certified operator for over 50 water systems in Arizona. To monitor fluoride reduction systems, we use a spectrophotometer and SPADNS reagent, which is simple, inexpensive, and can be performed in the field. Arsenic has no simple field test, but fortunately all of our arsenic-bearing waters also contain fluoride. We have found that fluoride breakthrough occurs before arsenic, and by testing for fluoride we can exchange units before arsenic breakthrough.

ACTIVATED ALUMINA

Activated alumina is primarily a hydrated aluminum oxide (Al_2O_3), which has been heated to a temperature of 300 to 700°C (570 to 1,290°F). It is then ground and screened to sizes ranging from 12.7-mm (0.5-in) granules to minus 325-mesh powders. The optimum size for POU water treatment applications has been 28 to 48 mesh (44 to 297 µm). AA particles are very irregular and porous with a very high surface area per unit mass. AA is an ion exchanger with the capability of exchanging both anions and cations. Alumina chromatography has been used successfully for separating organic as well as inorganic compounds. Acid-treated AA is primarily an anion exchanger with an anion selectivity sequence as follows:

OH^-, PO_3^{-3}, F^-, $Si(OH)_3O^-$, AsO_4^{-3}, $[Fe(CN)_6]^{-4}$, AsO_3^{-3}, CrO_4^{-2}, SO_4^{-2}, $Cr_2O_7^{-2}$, NO_2^{-1}, Br^{-1}, Cl^-, NO_3^-, MnO_4^-, ClO_4^-, CH_3COO^-

The anions are listed in their decreasing order of preference. The more preferred anions will tend to displace on the AA those anions that are lower in the sequence. Whereas in most organic anion exchangers, the fluoride ion is one of the least preferred, the reverse is true for AA. The OH^- ion is the most preferred and this has been a problem in POU applications. High alkalinity in water will reduce the capacity of AA to remove fluoride and arsenic. This problem is resolved in central plant fluoride reduction by lowering the pH of the water to be treated to approximately 5.5. Since such pH adjustment is not feasible with POU/POE applications, the lower capacity must be accepted.

Both of the forms of arsenic (AsO_4^{-3} arsenate and AsO_3^{-3} arsenite) found in water are in the anion selective sequence preferred to the SO_4^{-2} sulfate ion, and activated alumina has been used successfully as a POU/POE method for the removal of arsenic. Test facilities in Alaska, Oregon, and New Hampshire have determined that the activated alumina systems were very effective and much more economical then reverse osmosis and others tested.

PREPARATION

Activated alumina such as ALCOA F-1 or Kaiser Chemical A-2 should be pretreated before incorporation into a POU system. Pretreatment consists of a thorough backwash and acid wash. Aqueous state AA will have a pH of 9 to 10 and should be acid washed with a pH 2 sulfuric acid solution to a pH of 5 to 6. Backwashing is essential to remove the dust and fines in the material. If they are not removed, the alumina has a tendency to cement and destroy its adsorption capability. Activated alumina should always be added to an excess of water to dispell the heat generated, which will also contribute to cementing of the material.

CHEMISTRY

A simplified explanation of the absorption reactions (A = activated alumina particles):

1. ***Acid pretreatment***

$$A{\bullet}H_2O + H_2SO_4 \rightarrow A{\bullet}H_2SO_4 + H_2O$$
(aqueous state) (acid state)

2. ***Absorption or ion exchange***

$$A{\bullet}H_2SO_4 + 2NaF \rightarrow A{\bullet}2HF + Na_2SO_4$$

3. ***Regeneration***

$$A{\bullet}2HF + 4NaOH \rightarrow A{\bullet}2NaOH + 2NaF + 2H_2O$$

4. ***Neutralization***

$$A{\bullet}2NaOH + 2H_2SO_4 \rightarrow A{\bullet}H_2SO_4 + Na_2SO_4 + 2H_2O$$
(basic state)

REGENERATION

Several methods of regeneration of activated alumina have been explored over the past 15 years. Inconsistent results from regenerated media were a problem in many instances.

A regeneration method that gives uniformly good results has been recently developed. The process entails allowing complete replacement of the sorbate (Fl, AsO_4^{-3}, etc.) with OH^- and complete neutralization of the basic state AA with acid. Both of these reactions are time dependent, and allowing sufficient time for completion is the key to the success of this method. It has also been possible to greatly reduce chemical costs by recirculating the NaOH regenerant. It appears that the failure to completely neutralize the AA accounts for most of the loss of capacity on the media. In larger systems, where pH control is used, the continuing injection of acid is neutralizing the media well into the treatment cycle.

The steps in regeneration are:

1. *Backwash* - A backwash rate of 19.5 $m^3/m^2/h$ (8 gpm/sq ft) gives 100 percent expansion of the alumina bed, and the excess material is contained in an additional vessel (three to four minutes or until clear).

2. *Upflow regeneration* - One percent NaOH (by weight) is injected while the AA bed is still expanded. Regenerant is recycled for 30 minutes at 7.3 $m^3/m^2/h$ (3 gpm/sq ft).

3. *Soak* - Up to three hours.

4. *Neutralization* - Rinse with (pH = 2) dilute sulfuric acid until pH of alumina bed drops to 4.5-5.

5. *Soak* - 24 hours or until pH of alumina returns to 10^+ due to pore migration within the alumina particles.

Repeat steps 4 and 5 as needed to complete neutralizing of alumina bed.

6. *Final backwash and refill.*

CONCLUSION

Although the EPA has determined that POU technology is unacceptable for water treatment, we have successfully protected thousands of people from the damage caused by fluoride and arsenic in their drinking water. We have shown that POU treatment is a cost effective method of treatment in situations where the cost of central treatment would be prohibitive.

BIBLIOGRAPHY

1. Bellen, G.E., Anderson, M. and Gottler R.A. *Defluoridation of Drinking Water in Small Communities*. EPA/600/2-85/112, Cincinnati, OH, January 1986.

2. Bellack, E. 1971. Arsenic Removal from Potable Water. *JAWWA* 63:7:454, July 1971.

3. Clifford, D., Matson J., and Kennedy, R. Activated Alumina: Rediscovered "Adsorbent" for Fluoride, Humic Acids and Silica. *Industrial Water Engineering*, December 1978.

4. Harman, J.A. and Kalichman, S.G. Defluoridation of Drinking Water in Southern California. *JAWWA* 57:2:245, February 1965.

5. Kubli, H. On the Separation of Anions by Adsorption on Alumina. *Helvetia Chimica Acta*. Switzerland 3:453, 1947.

6. Maier, F.J. Defluoridation of Municipal Water Supplies. *JAWWA* 45:8:879, August 1953.

7. Savinelli, E.A. and Black, A.P. Defluoridation of Water with Activated Alumina. *JAWWA* 50::1: 33, January 1958.

8. Singh, G. and Clifford, D.A. *The Equilibrium Fluoride Capacity of Activated Alumina*. EPA-600/S2-81-082, Cincinnati, OH, July 1981.

MODELLING POINT-OF-ENTRY RADON REMOVAL BY GAC

Jerry D. Lowry
University of Maine
Orono, ME 04469

Sylvia B. Lowry
Lowry Engineering, Inc.
Thorndike, ME 04986

The health implications of airborne 222Rn in households are well documented (1-8), as is the importance of elevated 222Rn in the water supply and how it contributes to the airborne 222Rn (1,2,9-19).

The feasibility of removing 222Rn from household and small system water supplies with granular activated carbon (GAC) or aeration devices has been reported by various researchers (1,20-23) and a detailed report for public water supplies has recently been prepared for the U.S. Environmental Protection Agency (EPA) (24).

In response to the growing concern about airborne and waterborne 222Rn, research was initiated in 1980 at the University of Maine to identify technologies to remove 222Rn from ground water (17,22). Aeration and GAC were found to be potentially cost effective treatment processes for point-of-entry applications. Laboratory testing of full-scale point-of-entry GAC and diffused bubble aeration units showed that both methods were effective, but GAC appeared to have the most promise for household applications. That study and a previous one documented that an adsorption/decay steady state is established with a GAC bed, allowing it to be virtually maintenance free for an indefinite but long period of time. These studies led to laboratory research to model the GAC process for municipal application, field research to develop a design model for point-of-entry application, and installation and monitoring of the GAC technology in more than 100 households throughout the U.S. This article reports on the findings of the second and third aspects of the preceding research.

DEVELOPMENT OF GAC MODEL

Previous research indicated that the adsorption/decay steady state performance could be modelled by first-order kinetics, allowing the use of the following equation to describe and predict removal:

$$C_t = C_o e^{-K_{ss} t}$$

where:

- C_t = the 222Rn conc. at time t, pCi/l
- C_o = the initial 222Rn conc., pCi/l
- K_{ss} = the steady state adsorption/decay constant, hr^{-1}, and
- t = the empty bed detention time (EBDT), hr

This is logical since at steady state, the adsorptive removal equals the decay of adsorbed 222Rn at all points in the bed, and decay kinetics are first order. The actual achievement of the adsorption/decay steady state is quite complex. It involves nonsteady state adsorption kinetics, which dominate the early period, followed by an increasingly significant decay phenomenon set up by 222Rn and its short-lived progeny. The result of these processes is the establishment of the steady state distribution of 222Rn and short-lived progeny on the GAC. Although the establishment of the steady state is complex, the performance after steady state is reached can be accurately described by the simple steady state model.

The use of the average empty bed detention time (t) is appropriate since the GAC bed at steady state acts as a decay storage device. This means that for all practical purposes, the normal intermittent diurnal flow experienced in a typical household is not important, in terms of its effect on the steady state performance. An analogy to a chromatographic column can be made, in which the 222Rn travels along the bed at a relatively slow rate compared to the water, and decays down to the effluent value. The bed is simply a concentrating device that stores 222Rn and is equivalent to a plug flow storage tank having a much greater liquid detention time. For example, a GAC bed giving a 99 percent reduction is equivalent to an ideal plug flow decay storage tank having a detention time of 25.3 days. Thus, t should be calculated for the water volume used over two to three weeks because this is the period to which the GAC bed is responding. In a typical household the three week flow average does not vary significantly, except perhaps seasonally in a gradual manner.

To accurately document this model and to test the relative effectiveness of several different GAC products, a field study was designed to measure the steady state adsorption/decay constant (K_{SS}). A Maine household that had an extremely high 222Rn concentration in its ground water supply was selected to demonstrate the general applicability of the model across a wide range of 222Rn levels. The performance of three GAC products (Table 1) was examined using a modified commercial treatment unit (Figure 1). Slotted [(0.30 mm (0.01 inch)] laterals were installed at depths corresponding to 0.007, 0.014, 0.03, 0.04, 0.06 m^3 (0.25, 0.5, 1, 1.5, and 2 cu ft) bed volumes to obtain depth samples. The total bed volume was 0.07 m^3 (2.5 cu ft).

Table 1. Summary of Adsorption/Decay Design Constants

Carbon	Manufacturer	Type	K_{ss} (hr^{-1})
A	American Norit	Peat (8X20)	1.35
B	ICI Americas*	HD4000 (12X40)	2.09
C	Calgon	F-400 (12X40)	1.53
D	Barneby Cheney	299 or 1002	3.02

* Now manufactured by American Norit.

Water samples for 222Rn analysis were taken directly from sampling valves with a 10 ml (0.34 oz) syringe, which was subsequently discharged directly into a previously prepared liquid scintillation vial containing 5 ml (0.17 oz) of fluor. The vial was immediately capped, and mailed to the laboratory for counting. The basic counting procedure used was one described by Prichard and Gesell (25), except that a mineral oil- rather than toluene-based fluor was utilized because of postal regulations. Precision of the 222Rn analysis is a function of the level of 222Rn present, the counting time, and the time elapsed between sampling and counting. Typical levels of uncertainty (2-sigma) for this study are given in Table 2.

The water use at the household was monitored by a standard 16 mm (5/8 in) meter and totalizer readings were taken daily. Water temperature ranged between 6 and 10°C (43 and 50°F) throughout the study.

The various GAC products were tested sequentially, each by the same method. The virgin carbon was placed in the pressure vessel over a gravel support and commissioned after a backwashing period of 15 to 30 minutes to remove fines. The GAC bed remained in service to allow a steady state to be achieved (approximately three weeks) and was monitored for an additional three to four week period. Fourteen sets of samples (all ports) were taken to determine K_{SS}.

Figure 1. Experimental GAC vessel for K_{ss} determination.

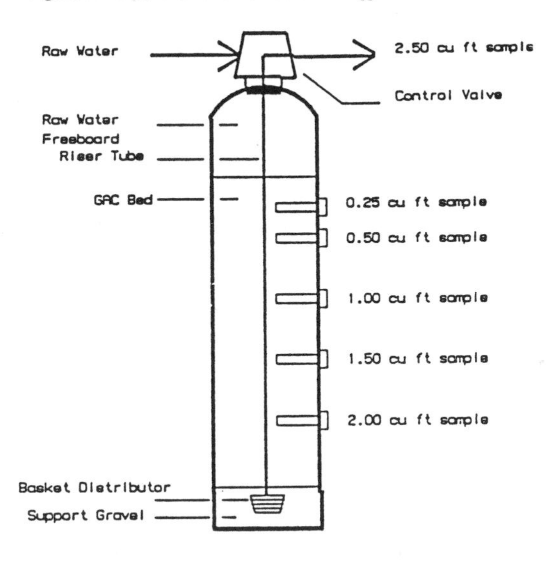

Table 2. Typical Levels of Uncertainty (Counting) for This Study

222Rn Concentration (pCi/l)	Uncertainty (percent)
800,000	0.5
300,000	0.5
40,000	1.0
2,000	4.5
1,000	6.0
500	15.0
100	25.0
60	45.0

Typical examples of the results of the field testing are illustrated by Figures 2 through 4 for GAC B. Figures 2 and 3 show results of the depth removal of 222Rn and the establishment of the adsorption/decay steady state. The exact reason for the elevated point for each depth on day 42 was not known, but suspected to be caused by desorption brought about by possible extreme raw water 222Rn variation that was not documented by sampling the previous week. This particular well is subject to such variations, and previous monitoring has documented that the 222Rn

Figure 2. Performance for the top portion of the GAC bed for GAC B.

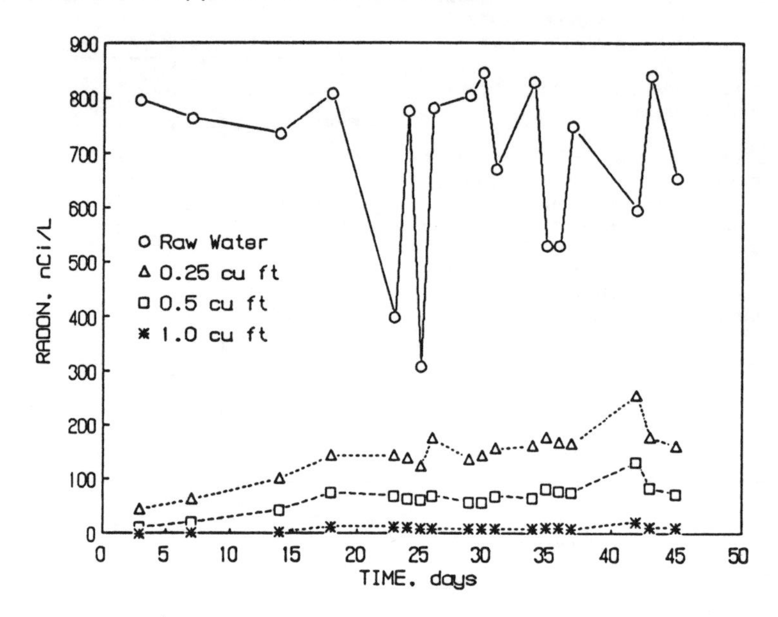

Figure 3. Performance for the bottom portion of the GAC bed for GAC B.

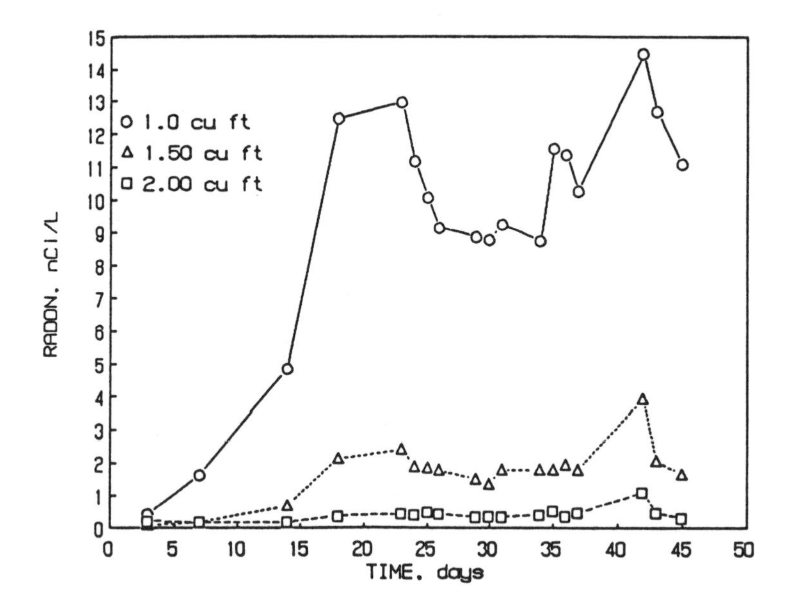

Figure 4. First-order steady-state adsorption-decay relation for GAC B, with 95 percent confidence limits indicated.

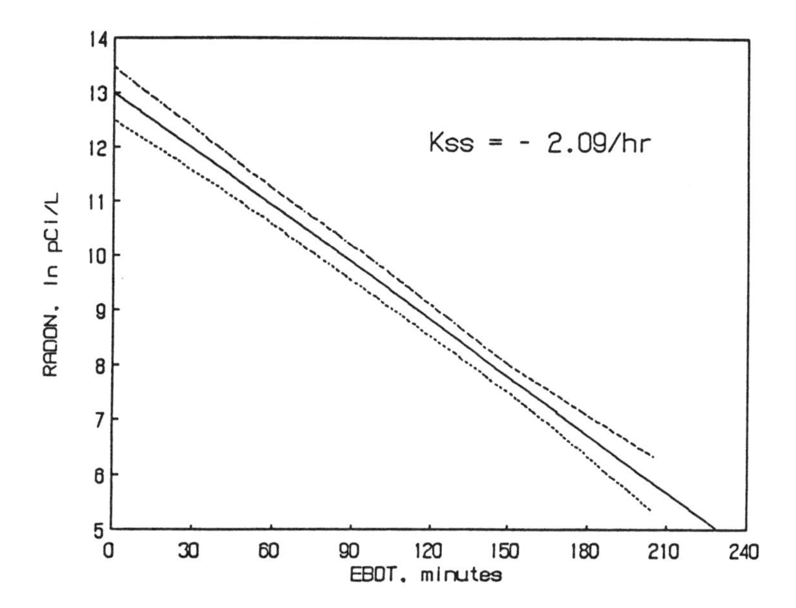

Figure 5. Contrast of the steady-state adsorption-decay relation for test carbons A, B, and C with GAC D.

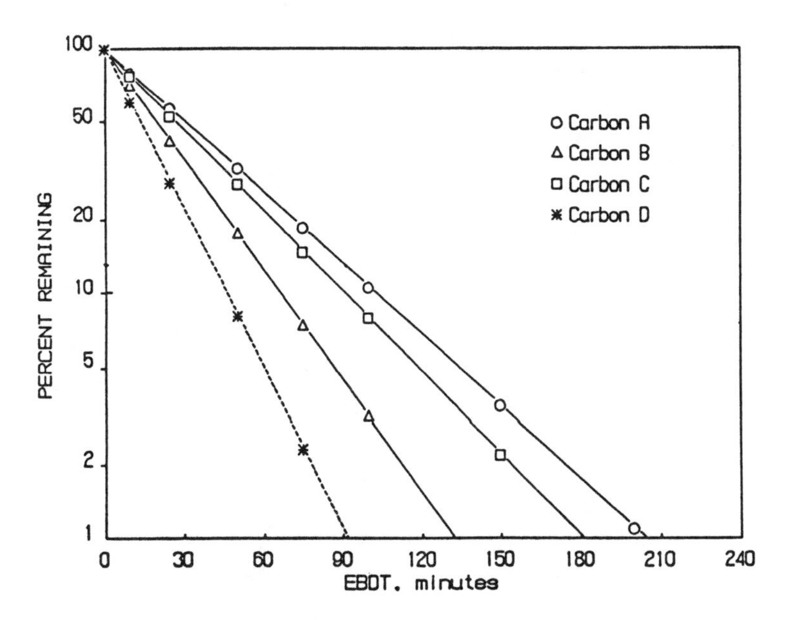

variation over a period as short as several days can be from 150,000 to over 2,000,000 pCi/l. For bed volumes of greater than 42 l (1.5 cu ft), these variations in depth are not significant in relation to the raw water concentrations because there is enough GAC to provide adequate dampening. They are apparent in this case due to the extremely high average raw water 222Rn.

A semi-ln plot of bulk solution 222Rn (ln vs. EBDT) yields a linear relationship with a slope equal to Kss. This relationship is illustrated in Figure 5 for one of the carbons tested.

The analysis of variance (ANOVA) for the data is summarized in Table 3. Although the deviations from regression were significant statistically, they were extremely small compared to the variation explained by regression. This fact is reflected by the relatively narrow confidence limits (95 percent) around the least squares regression line in Figure 4.

Table 3. Summary of Analysis of Variance for GAC B

Source of Variation	df	SS	MS	F
Among Groups (EBDT)	5	534.7	106.9	829*
Linear Regression	1	530.6	530.6	515*
Deviations from Regression	4	4.1	1.0	8.0*
Within Groups (ports)	78	10.1	0.13	

*Significant at 0.001 level.

The performance of carbons A, B, and C is contrasted to that of carbon D, which has been determined by other research (23,26) to have the highest K_{SS} value tested to date. A summary of the K_{SS} values for these carbons is given in Table 1. The ranking of these carbons bears little relation to how they performed previously by isotherm testing, and shows that isotherms are not indicators of how a carbon will perform at steady state (17). This group of four carbons contains the best and worst carbons with respect to 222Rn removal; it is clear that the type of GAC selected has significant bearing upon the performance achieved. For example at 99 percent removal, the required carbon volume for GAC D is 50 percent of that needed using carbon A. For equal bed volumes, carbon D achieves a 99 percent reduction compared to an 88 percent reduction with carbon A. It is interesting to note that the bulk densities for carbons A and D are approximately 288 and 513 kg/m^3 (18 and 32 lb/cu ft), respectively, and that on a mass, rather than volume, basis all GAC types tested are much closer in performance. Because the number and size of vessels required is determined by the volumetric performance, this has little practical significance. Of more importance is the probable positive influence of decreasing particle size; however, no studies have documented the magnitude of this factor for the 222Rn steady state.

Because of the small flow treated and the relatively narrow range of bed sizes required to cover the entire range of 222Rn encountered in point-of-entry applications, the long EBDT is easily satisfied by commercially available pressure vessels. A range of bed volumes from 28 to 85 l (1 to 3 cu ft) will remove in excess of 99 percent of the 222Rn from any household ground water supply. In contrast, the relatively long EBDT is of importance with larger design flows in municipal applications. Compared to an EBDT of approximately 15 minutes for organics removal, the required EBDT for 222Rn removal is quite long. Although the GAC will last indefinitely in the 222Rn application, the initial high capital cost for the GAC makes aeration an attractive alternative for larger water systems.

FIELD EXPERIENCE WITH GAC TREATMENT

Since 1981, GAC units have been installed in a significant number of households to remove 222Rn from the water supply and thereby lower the airborne 222Rn levels in homes. The current number of units that exist is estimated to be in excess of 500. Approximately 100 units have been installed and monitored as a part of a data base for future research on aspects other than simple removal, such as the resulting gamma exposure rate from 214Pb and 214Bi and the long term buildup of 210Pb. In each of these installations the GAC type, the GAC quantity, and the installation date are known. In some of the installations the water use is known and the radium and uranium content of the raw water has been documented. This data base is unique in that the units are installed over a widespread area (Canada and 12 states in the U.S) and contain elements that represent the longest operating GAC units for 222Rn. In addition, they are installed on water supplies that cover the entire documented range of 222Rn in the world -- from less than 1,000 pCi/l to in excess of 1,000,000 pCi/l.

Although four GAC types have been used in these units, over 85 percent contain carbon D and 10 percent contain carbon C. A summary of the steady state performance of all installations that are routinely monitored is presented in Figure 6. With the exception of three units, the performance level in field installations is very high. Eighty percent of all units are in the 0.05 m^3 (1.7 cu ft) category, with remaining units ranging from 0.03 to 0.08 m^3 (1 to 3 cu ft). The average removal of 222Rn for all units is 96.2 percent.

Elimination of three known prematurely fouled or malfunctioning units and the units containing carbon C, yields the histogram summarized in Figure 7. For these units, the average removal of 222Rn is 98.9 percent, and it is uncommon to monitor a unit and find less than 99 percent removal. Although the real

Figure 6. Histogram of the steady-state performance of 66 GAC treatment units.

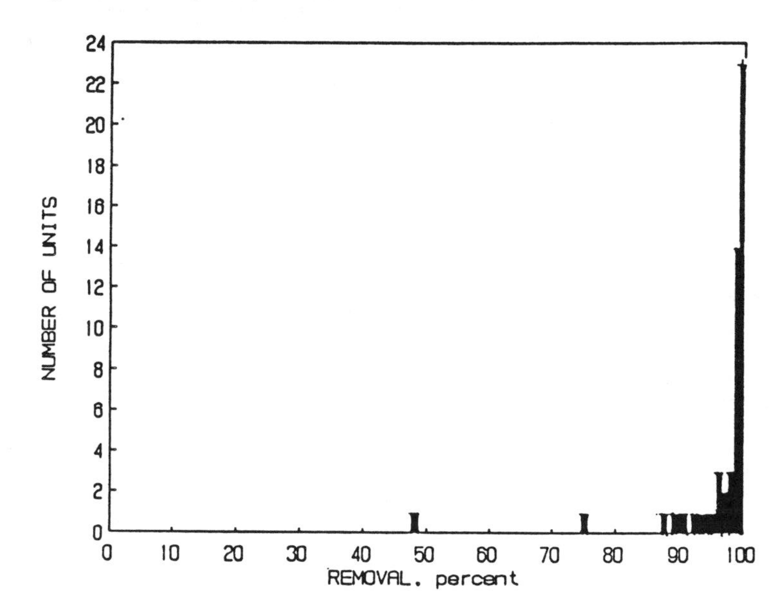

Figure 7. Histogram of the steady-state performance of properly operating GAC units conatining GAC D.

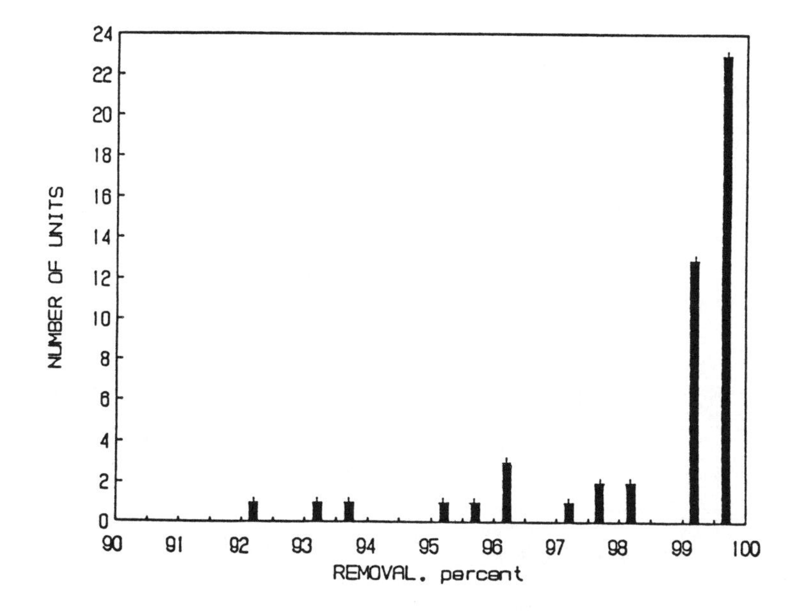

value of these units lies in future studies involving 222Rn progeny buildup, these data demonstrate the high degree of removal possible with properly designed and installed systems.

The reason for the three poorly performing units has not been determined. Preliminary investigation appears to indicate water quality as a cause, rather than the GAC. Although only a few percent of the existing GAC units exhibit this phenomenon, the reason for this possible premature fouling should be determined.

GAC vs. AERATION

It has been documented that aeration is a feasible method for 222Rn removal in point-of-entry applications (22,27). But a number of factors have kept it from becoming as popular a method as GAC treatment:

- Aeration is performed at atmospheric pressure and, therefore requires re-pressurization of the water supply.

- The initial cost of aeration systems designed for 222Rn removal is relatively high, partly due to the re-pressurization requirement. The installed cost ranges from $1,700 to over $4,000, compared to approximately $650 to $1,200 for GAC.

- Several of the currently available aeration methods have limited removal capabilities. A novel but costly spray aeration system was developed by the Maine Department of Human Services, Division of Health Engineering, and six units are operating in the field. These units achieve 90 to 95 percent removal but may be limited to wells containing only 10,000 to 20,000 pCi/l if, for example, the future U.S. EPA maximum contaminant level (MCL) for 222Rn is set at 1,000 pCi/l. Although the MCL would apply only to public water supplies, the real 222Rn issue is in private household supplies -- in many cases lending institutions are already requiring removal. These institutions tend to use the MCLs as guidance for supplies as well.

  Packed-tower aeration systems are confined by the ceiling height in the cellar or living area and are therefore limited to about 85 to 90 percent removal. Two such devices are currently available for household use. These units cost approximately $3,000.

  A multi-staged diffused bubble aeration system developed for organics removal (28) has been tested on a supply that contains 250,000 pCi/l and removed 222Rn to below detection, for virtually 100 percent removal. A less expensive version designed specifically for 222Rn removal has been developed that will achieve 99 percent removal. It will cost approximately $1,700.

- Aeration methods require significant O&M compared to GAC, which increases the cost differential over the long term.

Aeration methods have two advantages to be considered for point-of-entry applications, i.e., they avoid the elevated gamma exposure rate and long term buildup of 210Pb that is and may be, respectively, associated with GAC beds. These topics are currently the subjects of on-going research on GAC treatment and are beyond the scope of this paper; however, economical water jacket shielding and proper location can minimize increased gamma exposure. The buildup of 210Pb in these applications is not well documented, but it could be a concern from a regulatory point of view. Future documentation and research on this subject will determine the extent of and solutions to these problems.

CONCLUSIONS

- A first order model accurately describes the adsorption/decay steady state removal of 222Rn by GAC.

- A single design constant, K_{SS}, can be used to rank a given GAC type for 222Rn removal. The ranking of a carbon for steady state performance does not appear to be related to its ranking according to an adsorption isotherm.

- There is a significant range of the design constant, K_{SS}, for the carbons tested to date, making the selection of the correct GAC important. This is especially true for small public water supply application, where economics are more sensitive to vessel size.

- More than 99 percent reduction of 222Rn is possible with an effective GAC.

- The progeny of 222Rn make it important to consider the location of and protective shielding for a GAC bed, to minimize increased gamma exposure over background levels.

- Limited data indicate a possible premature fouling of approximately three to five percent of existing GAC units. The reason for this decreased removal in these installations is unknown and should be investigated.

ACKNOWLEDGEMENT

This paper is largely reproduced from a previously published paper in the Journal AWWA (October 1987).

Primary funding for the field research to determine the steady state design model was provided by the Office of Research and Development, U.S. Environmental Protection Agency (EPA), under Grant No. R8108290. The field monitoring data was supplied by Lowry Engineering, Inc.

REFERENCES

1. Castren, O. The contribution of bored wells to respiratory radon daughter exposure in Finland. Technical Report, Institute of Radiation Protection, Helsinki, Finland, 1977.

2. Hess, C. T. et al. Radon in potable water supplies in Maine: the geology, hydrology, physics, and health effects. Completion Report, Land and Water Resources Center, University of Maine and the Office of Water Research and Technology, U.S. Department of the Interior, Washington, DC, 1979.

3. Evans, R. D. et al. Estimate of risk from environmental exposure to radon-222 and its decay products. Nature. Vol. 290, 1981, pp. 98-100.

4. Harley, N. H. Editorial--radon and lung cancer in mines and homes. New England Journal of Medicine. Vol. 310, No. 23, June 7, 1984, pp. 1525-1526.

5. National Academy of Sciences Committee on Biological Effects of Ionizing Radiation (BEIR). The effects on populations of exposure to low levels of ionizing radiation. National Academy Press, Washington, DC, 1980.

6. U.S. Environmental Protection Agency. A citizens guide to radon: what it is and what to do about it. August, 1986.

7. U.S. Environmental Protection Agency. Radon reduction methods: a homeowners guide. U.S. EPA OPA-86-005, August, 1986.

8. Cothern, R. C. Estimating the health risks of Radon in drinking water. Journal of the American Water Works Asso. Vol. 79, No. 4, April, 1987.

9. Castren, O. et al. High natural radioactivity of bored wells as a radiation hygienic problem in Finland. Presented at the 1977 International Radiation Protection Association Fourth International Congress, Paris, France.

10. Duncan, D. L., Gesell, T. F. and Johnson, R. H. Radon-222 in potable water. Proceedings of the 10th Midyear Topical Symposium: Natural Radioactivity in Man's Environment, Health Physics Society, Saratoga Springs, NY, October, 1976.

11. Hess, C. T. et al. Investigation of Rn-222, Ra-226, and U in air and groundwater of Maine. Completion Report B-017-ME, Land and Water Resources Center, University of Maine and the Office of Water Research and Technology, U.S. Department of the Interior, Washington, DC, 1981.

12. Partridge, J. E., Horton, T. R. and Sensintaffer, E. L. A study of radon-222 released during typical household activities. ORP/EERF-79-1, U.S. Environmental Protection Agency, Office of Radiation Programs, Eastern Environmental Radiation Facility, Montgomery, AL, March, 1979.

13. Weiffenbach, C. V. Radon in air and water: health risks and control measures. Technology Transfer Report, Land and Water Resources Center, University of Maine, June, 1986.

14. Horton, T. R. Methods and results of EPA's study of radon in drinking water. EPA 520/5-83-027, U.S. Environmental Protection Agency, Office of Radiation Protection Programs-Eastern Environmental Radiation Facility, Montgomery, AL, December, 1983.

15. Aldrich, L. K., Sasser, M. K. and Conners, D. A. Evaluations of radon concentrations in North Carolina ground water supplies. Department of Human Resources Division of Facility Services-Radiation Protection Branch, Raleigh, NC, January, 1975.

16. Lowry, J. D. and Moreau, E. Removal of extreme radon and uranium from a water supply. Proceedings of the 1986 National Conference on Environmental Engineering, Environmental Engineering Division, American Society of Civil Engineers, Cincinnati, OH, July, 1986.

17. Lowry, J. D. and Brandow, J. E. Removal of radon from water supplies. Journal of Environmental Engineering, Environmental Engineering Division, American Society of Civil Engineers. Vol. 111, No. 4, August, 1985.

18. Lowry, J. D. Radon at home. Civil Engineering. Vol. 57, No. 2, February, 1987.

19. Prichard, H. M. The transfer of radon from domestic water to indoor air. Journal of the American Water Works Asso. Vol. 79, No. 4, April, 1987.

20. Hoather, R. C. and Rackham, R. F. Some observations on radon in waters and its removal by aeration. Proceedings of the Institution of Civil Engineers, Great George Street, London, S.W.1., December 7, 1962, pp. 13-22.

21. Smith, B. M. et al. Natural radioactivity in ground water supplies in Maine and New Hampshire. Journal of the American Water Works Association. Vol. 53, No. 1, January, 1961, pp. 75-88.

22. Lowry, J. D. et al. Point-of-entry removal of radon from drinking water. Journal of the American Water Works Asso. Vol. 79, No. 4, April, 1987.

23. Lowry, J. D. Extreme levels of Rn-222 and U in a private water supply. Proceedings of the Conference on Radon, Radium, and Other Radioactivity in Ground Water: Hydrologic Impact and Application to Indoor Airborne Contamination, National Water Well Asso., Somerset, NJ, April 7-9, 1987.

24. Malcolm Pirnie, Inc. Technologies and costs for the removal of radon from potable water supplies. Draft of report to EPA under Contract No. 68-01-6989, January 6, 1987.

25. Pritchard, H. M. and Gesell, T. F. Rapid measurements of radon-222 concentrations in water with a commercial liquid scintillation counter. Health Physics. Vol. 33, 1977, pp. 577-581.

26. Pinnette, J. M.S. thesis in Civil Engineering, University of Maine, Orono, ME, 1985.

27. Hinkley, W. W. Experimental water treatment for a drilled well with the world's highest known radon-222 levels. Maine Dept. of Human Services, Div. of Health Engineering, State House, Augusta, ME, 1982.

28. Lowry, J. D. and Lowry, S. B. Restoration of gasoline-contaminated household water supplies to drinking water quality. Proceedings of the Eastern Regional Ground Water Conference, National Water Well Asso., Portland, ME, July, 1985.

POINT-OF-ENTRY ACTIVATED CARBON TREATMENT LAKE CARMEL - PUTNAM COUNTY

George A. Stasko
Bureau of Public Water Supply Protection
NY State Department of Health
Albany, NY 12237

Lake Carmel is a small lake located approximately 80 km (50 mi) north of New York City. The hills surrounding the lake were extensively developed in the 1930s with seasonal residences. Since then most of these residences have been converted to year round housing. Lot sizes range from 370 to 1,110 m^2 (4,000 to 12,000 sq ft) and each lot has a well and septic system (see Figure 1).

In the 1970s, some of the residents complained about petroleum odors in their water. When individual action did not bring the desired results they formed a Citizens' Advisory Committee. The committee enlisted the aid of their legislators with the result that investigations of the area were made by the Department of Transportation, the State Health Department, and the Putnam County Health Department.

INVESTIGATION RESULTS

The residents believed that the ground water contamination came from a petroleum spill or from a waste site. However, as a result of the investigations, it was concluded that the residents had contaminated their own water supplies by localized petroleum leaks and spills, and by the chemicals flushed into their septic systems. In addition, it was discovered that some wells had elevated nitrate levels and high coliform counts.

BACTERIOLOGICAL QUALITY

Bacteriological test results indicated widespread bacterial contamination. Approximately 40 percent of the samples were above the standard of 1 coliform organism per 100 ml (0.3 organism per oz). Counts as high as 245 coliform organisms per 100 ml (72 per oz) were found (see Table 1).

Coliform levels varied considerably from well to well and in the same well over a period of time. In the Fitzsimmons well, 245 coliform organisms per 100 ml (72 per oz) were present on February 4, 1982. Two weeks later the count was 3 per 100 ml (0.9 per oz) and four weeks after the second sample the count was 134 per 100 ml (39 per oz). In the Saver well on February 18, 1982 the count was 189 coliform organisms per 100 ml (56 per oz), and on March 23, 1982 the count was less than 1 coliform organism per 100 ml (0.3 per oz). Because of this variability, the health department concluded that all wells would be subject to bacterial contamination at some point in time and therefore the treatment system must include disinfection.

ORGANIC CHEMICAL QUALITY

Volatile Organic Chemicals

Detection of volatile organic chemicals varied from well to well, and with the laboratory performing the analyses. All the highest results were obtained from one laboratory. Other laboratories detected the same chemicals but at lower levels. Sampling for the three studies was not coordinated, leaving gaps in the data.

Benzene, toluene, and xylene were detected at high levels confirming residents' complaints of petroleum tastes and odors. In addition, several solvents were found including carbon tetrachloride, tetrachloroethylene, trichloroethylene, and 1,1,1-trichloroethane. Table 2 summarizes the results of tests for volatile organic chemicals.

Base/Neutral Chemicals

Of the base/neutral organic chemicals tested, only bis (2-ethylhexyl) phthalate was found in detectable quantities (see Table 3).

Pesticides/Herbicides

No pesticides or herbicides were detected. For the chemicals tested see Table 4.

INORGANIC CHEMICAL QUALITY

The average inorganic water quality of the wells tested was soft and corrosive, but no chemicals exceeded standards. Both sodium and nitrates were significantly higher than background levels in area ground water, indicating that leachate from the septic systems was reaching the wells. Only in one instance did nitrate exceed the standard of 10 mg/l (10 ppm). It

Figure 1. Location map.

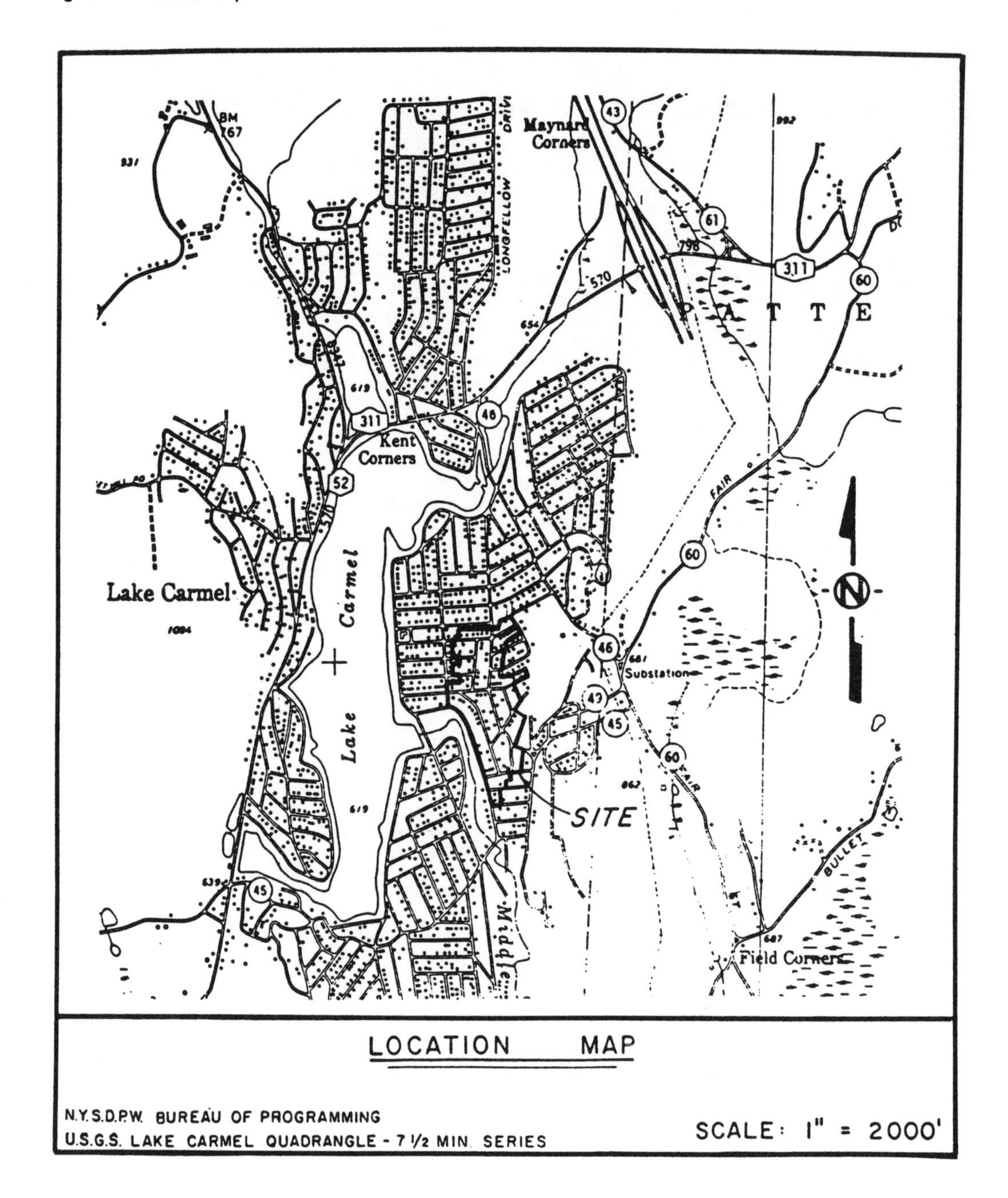

Table 1. Bacteriological Quality of the Ground Water in the Lake Carmel Project Area*

Name	Collection	Total Coliform per 100 ml	Above Health Limits**
Backer	3/24/82	57	Yes
Baisley	4/15/82	< 1	No
Bao	4/27/82	< 1	No
Behnken	2/18/82	6	Yes
Cuomo	4/21/82	< 1	No
Fitzsimmons	2/4/82	245	Yes
Fitzsimmons	2/18/82	3	Yes
Fitzsimmons	3/16/82	134	Yes
Greco	2/4/82	< 1	No
Laconte	2/4/82	91	Yes
Laconte	3/16/82	17	Yes
Lawton	3/24/82	92	Yes
MacNeil	3/23/82	< 1	No
Madden	4/15/82	< 1	No
Mahoney	2/4/82	< 1	No
Micciche	4/21/82	< 1	No
Nappi	4/21/82	< 1	No
Placek	1/5/82	38	Yes
Porrino	3/23/82	< 1	No
Prisco	4/15/82	< 1	No
Sauer	2/18/82	189	Yes
Sauer	3/23/82	< 1	No
Sheridan	2/4/82	< 1	No
Sutor	2/4/82	< 1	No
Zasso	3/18/82	< 1	Yes
Zasso	4/21/82	< 1	No

* All samples were collected by the Putnam County Department of Health and analyzed by Sanitary Science & Laboratories, Inc., Newburgh, New York.
** The NYS Dept. of Health limits total coliforms to 1/100 ml.

Table 2. Organic Chemical Quality of the Ground Water in the Lake Carmel Area (Volatile Compounds)

Volatile Compound	High Value (μg/l)	Mean Value* ± Std. Dev. (μg/l)
Acrolein	ND	ND
Acrylonitrile	ND	ND
Benzene	820	67.7 ± 176.3 (21)
Bromodichloromethane	4	2.90 ± 1.15 (3)
Bromoform	ND	ND
Bromomethane	ND	ND
carbon Tetrachloride	5.9	4.50 ± 1.98 (5)
Chlorobenzene	ND	ND
Chlorodibromomethane	ND	ND
Chloroethane	ND	ND
2-Chloroethyl vinyl ether	ND	ND
Chloroform	14	5.03 ± 4.81 (8)
Chloromethane	ND	ND
Dichlorodifluoromethane	ND	ND
1,1-Dichloroethane	ND	ND
1,2-Dichloroethane	ND	ND
1,1-Dichloroethylene	ND	ND
Trans-1,2-Dichloroethylene	ND	ND
1,2-Dichloropropane	ND	ND
1,3-Dichloropropane	ND	ND
Ethylbenzene	110	38.3 ± 62.1 (3)
Methylene chloride	ND	ND
1,1,2,2-Tetrachloroethane	ND	ND
Tetrachloroethylene	560	100.2 ± 215.3 (11)
Toluene	320	8.5 ± 88.0 (13)
1,1,1-Trichloroethane	14.2	6.91 ± 5.10 (9)
1,1,2-Trichloroethane	ND	ND
Trichloroethylene	13	5.48 ± 3.86 (10)
Trichlorofluoromethane	2	1.25 ± 0.50 (4)
Vinyl chloride	2	2 ± 0 (2)
Xylene	490	108.4 ± 178.1 (7)

ND - Not detectable.
* Numbers in parentheses indicate sample size.

was decided that no treatment was needed to remove inorganic chemicals. See Table 5 for a summary of inorganic chemical test results.

Alternate Solutions

As a result of the investigations, the health department recommended that a public water system be provided for the affected area. An engineering consultant was hired to conduct a feasibility study. The study found that a public water system would cost in excess of $1,200 per year per homeowner. Expensive rock cut for the distribution system was the main reason why the public water system would be so costly. Because the public water system was impractical, it was decided to study the feasibility of a point-of-entry solution.

The Citizens' Advisory Committee was able to secure an imminent threat grant from the U.S. Department of Housing and Urban Development. They provided $165,000 to design, purchase, and install point-of-entry treatment systems. Because the Citizens' Advisory Committee could not receive the money, it was given to the town of Kent, which in turn hired an engineering firm to design the treatment systems.

Water Treatment Systems

The health department worked closely with the consultants to develop a water treatment system that would adequately treat the water and would satisfy all regulatory concerns. The design that resulted evolved from design criteria developed by the health department's Ad Hoc Committee on Removal of Synthetic Organic Chemicals from Drinking Water, and published in an interim report, entitled *Point-of-Use Activated Carbon Treatment Systems.*

A schematic of the treatment system is shown in Figure 2. The treatment system consists of:

- A raw water tap located immediately after the homeowner's pressure tank used to collect untreated water samples.

Table 3. Organic Chemical Quality of the Ground Water in the Lake Carmel Area (Base/Neutral Compounds)

Base/Neutral Compound	High Value (µg/l)	Mean Value* ± Std. Dev. (µg/l)
Acenaphthene	ND	ND
Acenaphthylene	ND	ND
Anthracene	ND	ND
Benzo (a) anthracene	ND	ND
Benzo (b) fluoroanthene	ND	ND
Benzo (k) fluoroanthene	ND	ND
Benzo (a) pyrene	ND	ND
Benzo (g,h,i) perylene	ND	ND
Benzidine	ND	ND
Bis (2-chloroethyl) ether	ND	ND
Bis (2-chloroethoxy) methane	ND	ND
Bis (2-ethylhexyl) phthalate	35	23 ± 17.0 (2)
Bis (2-chloroisopropyl) ether	ND	ND
4-Bromophenyl phenyl ether	ND	ND
Butylbenzylphthalate	ND	ND
2-Chloronaphthalene	ND	ND
4-Chlorophenylphenylether	ND	ND
Chrysene	ND	ND
Dibenzo (a,h) anthracene	ND	ND
Di-N-Butylphthalate	ND	ND
1,2-Dichlorobenzene	ND	ND
1,3-Dichlorobenzene	ND	ND
1,4-Dichlorobenzene	ND	ND
3,3'-Dichlorobenzidine	ND	ND
Diethylphthalate	ND	ND
Dimethylphthalate	ND	ND
2,4-Dinitrotoluene	ND	ND
2,6-Dinitrotoluene	ND	ND
Di-octyl-phthalate	ND	ND
1,2-Diphenylhydrazine	ND	ND
Fluoroanthene	ND	ND
Fluorene	ND	ND
Hexachlorobenzene	ND	ND
Hexchlorobutadiene	ND	ND
Hexchloroethane	ND	ND
Hexachlorocyclopentadiene	ND	ND
Indeno (1,2,3-cd) pyrene	ND	ND
Isophorone	ND	ND
Naphthalene	ND	ND
Nitrobenzene	ND	ND
N-Nitrosodimethylamine	ND	ND
N-Nitrosodi-N-propylamine	ND	ND
N-Nitrosodiphenylamine	ND	ND
Phenanthrene	ND	ND
Pyrene	ND	ND
1,2,4-Trichlorobenzene	ND	ND
2,3,7,8-Tetrachlorodibenzo-p-dioxin	ND	ND

ND - Not detectable.
* Numbers in parentheses indicate sample size.

- A water meter to measure the amount of water processed.

Table 4. Organic Chemical Quality of the Ground Water in the Lake Carmel Area

Pesticide/Herbicide	High Value (µg/l)	Mean Value ± Std. Dev. (µg/l)
Aldrin	ND	ND
-BHC		
-BHC		
-BHC		
-BHC		
Chlordane	ND	ND
Dieldrin	ND	ND
-Endosulfan		
-Endosulfan		
Endosulfan sulfate	ND	ND
Endrin	ND	ND
Endrin aldehyde	ND	ND
Heptachlor	ND	ND
Heptachlor Epoxide	ND	ND
4,4'-DDT	ND	ND
4,4'-DDE	ND	ND
4,4'-DDD	ND	ND
PCB 1016	ND	ND
PCB 1221	ND	ND
PCB 1232	ND	ND
PCB 1242	ND	ND
PCB 1248	ND	ND
PCB 1254	ND	ND
PCB 1260	ND	ND
Toxaphene	ND	ND

ND - Not detectable.

- Two 5-µm (0.0002-in) cartridge-type prefilters in parallel to prevent the activated carbon filters from clogging due to particulate matter. Backwashing of the activated carbon filters is not recommended because of the operational problems this would cause, due to the difficulty of disposing of the backwash water and the difficulty in obtaining enough treated water at adequate pressure to provide an adequate backwash.

- Two activated carbon filters in series. Each filter consists of 25.4-cm (10-in) diameter fiberglass tank containing 18 kg (40 lb) of virgin activated carbon. Bed depth is 91 cm (36 in) and each cylinder has a empty bed contact time of approximately 2.5 minutes at a flow rate of 0.32 l/s (5 gpm). The theoretical lifetime of this treatment system is 36 months based on an influent Benzene concentration of 244 µg/l. Benzene at this level was chosen as the critical design factor because it will give a 99 percent assurance that the treatment system would meet any influent organic chemical challenge encountered during the testing program. The filters are operated in series with the lead cylinder changed yearly. This is done, even though the theoretical lifetime is 18 months, to provide a factor of safety. When the lead cylinder is removed

Table 5. Inorganic Chemical Quality of the Ground Water in the Lake Carmel Area

Element	High Value (µg/l)	Mean Value* ± Std. Dev. (µg/l)
Aluminum (Al)	0.326	0.095 ± 0.155 (4)
Arsenic (As)	0.060	0.015 ± 0.030 (4)
Barium (Ba)	0.290	0.115 ± 0.122 (4)
Beryllium (Be)	0	0 ± 0 (4)
Boron (B)	0.849	0.212 ± 0.425 (4)
Cadmium (Cd)	0	0 ± 0 (4)
Calcium (Ca)	56.81	30.27 ± 16.04 (12)
Chromium (Cr)	0.005	0.001 ± 0.002 (4)
Cobalt (Co)	0	0 ± 0 (4)
Copper (Cu)	0.078	0.033 ± 0.039 (4)
Iron (Fe)	0.268	0.100 ± 0.097 (12)
Lead (Pb)	0.030	0.008 ± 0.015 (4)
Magnesium (Mg)	15.68	10.47 ± 5.45 (12)
Manganese (Mn)	0.372	0.081 ± 0.126 (12)
Molybdenum (Mo)	0.014	0.005 ± 0.007 (4)
Nickel (Ni)	0.009	0.002 ± 0.005 (4)
Phosphorous (P)	0.095	0.063 ± 0.043 (4)
Potassium (K)	4.29	2.18 ± 2.42 (4)
Selenium (Se)	0	0 ± 0 (4)
Silicon (Si)	8.12	4.69 ± 2.75 (4)
Silver (Ag)	0	0 ± 0 (4)
Sodium (Na)	169.7	85.28 ± 73.49 (4)
Vanadium (V)	0	0 ± 0 (4)
Zinc (Zn)	0.068	0.033 ± 0.038 (4)
Nitrate	13.67	5.65 ± 3.47 (23)

* Numbers in parentheses indicate sample size.

the lag cylinder is moved to the lead position and the new cylinder is placed in the lag position.

- A valving arrangement is provided to allow for water use during the cylinder changing procedure.
- Pressure gauges are provided before and after the treatment system to determine head loss across the system.
- Ultraviolet light disinfection is provided after the activated carbon units to destroy bacteria that break through the filter system. A light sensor with a visual alarm is provided on the ultraviolet light unit to inform the homeowner of proper disinfection.

WATER TREATMENT SYSTEM COST

To receive as many bids as possible, the engineer arranged for the system to be bid in six separate contracts. Low bid results are listed in Table 6.

WATER SYSTEM MANAGEMENT

The original concept for management of the water treatment systems was for the Town of Kent to own the systems and to be responsible for their operation and maintenance. This responsibility for maintenance could be carried out by their own personnel or under contract by a qualified agent. The County and State Health Departments would provide technical assistance and some monitoring and regulatory oversight.

However, the town turned the responsibility for operation and maintenance over to the homeowners. They formed a not-for-profit corporation, the Lake Carmel Water Quality Improvement District (LCWQID). The corporation consists of all the home owners who have treatment systems. The homeowners elect a President, Vice-President, Secretary, Treasurer, and a seven member Board of Directors. The Board of Directors consists of the Officers of the Corporation, who were also the active members of the Citizens' Advisory Committee, in addition to three maintenance men.

Participation in the district is voluntary. Sixty-seven of the 110 eligible homes received treatment systems.

OPERATION AND MAINTENANCE

Operation and maintenance consists of changing one of the activated carbon cylinders each year and changing the bulb on the ultraviolet unit every nine months. At the beginning of each year, the home owner is given enough refill cartridges for the prefilters and is expected to change them when necessary.

Recharge of the activated carbon cylinders is accomplished by taking the spent cylinder to a town-provided workshed. The used activated carbon is emptied and the fiberglass cylinder is refilled with a bed of sand and 18 kg (40 lb) of virgin activated carbon. The spent carbon is disposed of at a landfill.

Maintenance men make house calls to repair leaks and to clean the quartz tube on the ultraviolet unit. They are paid on a flat rate per item basis.

ANNUAL COSTS

For the first four years of operation the annual cost for operation, maintenance, and monitoring has been $250 per treatment system. The annual charge has recently been raised to $320 per year paid on a quarterly basis. This amounts to a total annual budget of $20,480 for the district.

WATER TREATMENT SYSTEM PERFORMANCE

Although there is no legal requirement for the district to monitor and report the performance of the water treatment systems, they have tried to follow the

Figure 2. Water treatment system.

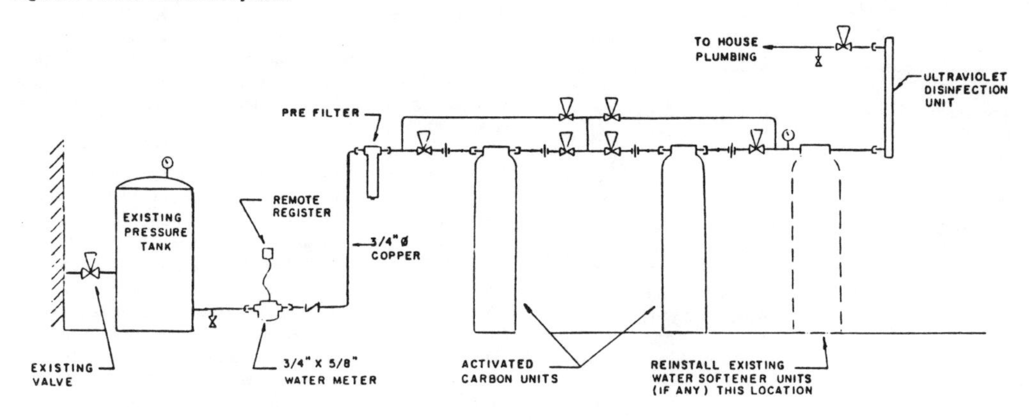

Table 6. Water Treatment System Cost

Item	Cost ($)
1 Water Meter 8 Gate Valves 1 Check Valve 3 Sampling Taps 2 Pressure Gauges	150.60
2 Cartridge Filter Units	67.84
2 Fiberglass Cylinders	140.60
80 Pounds Activated Carbon @ $0.90/lb	72.00
1 Ultraviolet Disinfection Unit	392.00
Installation	494.00
Total System Cost	1,317.04

guidance given them to sample at least 10 percent of the systems each year.

BACTERIOLOGICAL PERFORMANCE

During the years of 1984, 1985, and 1986, 21 paired samples were collected and analyzed for coliform organisms. Of these, three untreated samples had high coliform counts and one treated sample had a count of 2 coliform organisms per 100 ml (0.6 per oz). This is a great improvement over the original sampling where 40 percent of the drinking water samples had high coliform counts.

ORGANIC PERFORMANCE

The district has not had enough money to do adequate organic chemical monitoring. In 1984, 10 paired samples were collected and analyzed for benzene, toluene, and xylene. The detection limit was too high to show system performance, but the sample results did indicate that these chemicals were not present in either the untreated or treated water above the guideline levels of 5 µg/l for benzene, and 50 µg/l for toluene and xylene. In 1985, four paired samples of these same chemicals were analyzed with all results below 1 µg/l.

In 1986 and 1987, the New York State (NYS) Department of Transportation collected a series of volatile organic samples at one residence because of a nearby chemical spill. On December 23, 1986, samples were collected before and after the treatment system. The results are shown in Table 7.

Table 7. NYS DOT Volatile Organic Samples: December, 1986

Chemical	Before Treatmen (µg/l)	AfterTreatmen (µg/l)
Toluene	5	<1
Ethylbenzene	3	<1
p-Xylene	2	<1
m-Xylene	6	<1
o-Xylene	4	<1
n-Propylbenzene	2	<1
1,3,5 Trimethylbenzene	4	<1
1,2,4 Trimethylbenzene	4	<1
Cyclopropylbenzene	5	
Total	35	<1

On February 5, 1987, samples were collected before and after the treatment system. The results are shown in Table 8.

On March 7, 1987, no contaminants were detected in a treated sample, and on May 8, 1987 and August 12, 1987, no contaminants were detected in samples collected before and after treatment.

Table 8. NYS DOT Volatile Organic Samples: February, 1987

Chemical	Before Treatmen (µg/l)	AfterTreatmen (µg/l)
Toluene	5	5
m-Xylene	2	2
o-Xylene	1	1
o-Chlorotoluene	2	2
p-Chlorotoluene	1	<1
1,3,5 Trimethylbenzene	2	<1
Cyclopropylbenzene	3	2
o Dichlorobenzene	5	4
Hexachlorobutadiene	6	<5
Total	27	16

The results of all organic samples except one show removal of the tested organic chemicals to below detectable levels. In the sample collected on February 5, 1987, contaminants were detected in the treated water sample at 5 µg/l (5 ppb) or less. The limited test results are not comprehensive enough to make a definitive statement on the removal of organic contaminants, but they do give an indication that the treatment system is performing satisfactorily.

COMMUNITY DEMONSTRATION OF POU SYSTEMS REMOVAL OF ARSENIC AND FLUORIDE: SAN YSIDRO, NEW MEXICO

Karen Rogers
Leedshill-Herkenhoff, Inc.
Albuquerque, NM 87103

San Ysidro is a small, rural village of approximately 200 people located in the north central part of the State of New Mexico, approximately 50 minutes north of Albuquerque. The village is at least 200 years old. It was settled by Spanish colonists on a land grant from Spain in 1786. San Ysidro lies between the lands of the Zia and Jemez Pueblo Indians along the Jemez River. Life in the community is simple and relaxed and most residents live there for exactly that reason. The mean annual income for families in San Ysidro is $13,500. Fifty-eight percent of families earn less than $10,000 a year.

The village water supply is collected in an infiltration gallery into which ground water is drawn. This local ground water contains leachate from geothermal activity in the area's abundant mineral deposits and is therefore high in mineral content. The ground water exceeds the standards and/or maximum contaminant levels (MCLs) for arsenic, fluoride, iron, manganese, chloride, and total dissolved solids. The contaminants of concern in the village water supply are arsenic V and III and fluoride, which exceed the MCLs by three to four times (Table 1).

Table 1. Average Water Quality in San Ysidro, New Mexico

Contaminant	Conc. (mg/l)	Max. Contaminant Level or Rec. Std. (mg/l)	Avg. After RO (mg/l)
Iron	2.0	0.3	0.015
Manganese	0.2	0.05	<0.01
Chloride	325.0	250.0	12.50
Fluoride	5.2	1.8*	0.40
Arsenic V & III	0.22	0.05	<0.01
TDS	1,000.0	500.0	<180.0

* Recently revised to 4.0

Prior to discussing the point-of-use (POU) treatment that is installed in San Ysidro, a better understanding of the San Ysidro water system is necessary, including some general problems which have a direct bearing on the future success or failure of the point-of-use devices.

As mentioned before, the village water supply source is an infiltration gallery that produces an average of 27,000 gpd in winter and 36,000 gpd in summer from the ground water. The infiltration gallery has a storage capacity of 17,000 gallons. The village currently uses an average of 30,000 gpd. This equates to about 150 gpd per person. This consumption rate pushes the production/storage capacity limits of the gallery.

There is a 20,000-gal elevated storage tank connected into the piping system that should be providing the additional capacity the village needs, but it has seldom been a functioning unit for several reasons. First, the pumps that are currently located in the infiltration gallery do not have adequate controls to allow them to operate appropriately to maintain an adequate supply of water in the system. There is no remote readout on the status of the pumps or system. The only way to know there is a problem is when a faucet is opened and no water comes out. Secondly, the village does not have one person who knows the system and who has the responsibility to keep it operating. A village employee, the major or one of the village council usually goes down to turn the pumps on when someone calls to complain about the low pressure or to report that they have no water. The pumps should be running all night, but because of other problems with the controls and overheating pump motors, someone would need to monitor them all night to insure safe operation; however, this is not an acceptable solution. So, the village has a long history of water supply problems including low water pressure, no water at all, and quality problems including taste, color, clarity, and odor in addition to the contaminant levels discussed earlier.

A number of alternatives have recently been investigated by engineers at Leedshill-Herkenhoff to aid the village in obtaining a higher quantity and quality of water. When the village first was found to be in violation of the Safe Drinking Water Act (SDWA) for levels of arsenic and fluoride in the water supply, four deep test wells were drilled to determine if there was a better source of water available. The best of these wells only had water equal in quality to the water in the infiltration gallery. It was at this point that treatment options began to be studied. The recommendation for added system capacity at this point was to increase the size of the infiltration gallery. The village obtained bids to perform this work but all of them exceeded the available funds because of the high costs of dewatering the site during construction. The village then decided to add the best test well that had been drilled previously to the system. This new well was recently developed at 43,200 gpd and is being pumped into the infiltration gallery temporarily to supplement the system. The new well will be permanently tied into the system under a project which is currently being reviewed by the state for approval. This project will also provide an automatic control system to regulate pumping from the well or the gallery to maintain a beneficial water level in the village storage tank. This control system will have remote readouts in the village office to indicate if the system is functioning normally or to show there is a problem.

A variance from the SDWA for arsenic and fluoride was granted to the village while research was performed by Dennis Clifford of the University of Houston to determine an economical and effective solution to the contaminant problem. The treatment systems studied were activated alumina and reverse osmosis (RO). Central and point-of-use treatment were considered. Central treatment of the entire water supply was not considered feasible for many reasons. First, there is a disposal problem with both the arsenic-contaminated sludge from activated alumina column regeneration and the reject brine from the reverse osmosis unit. Secondly, the costs of central treatment were considered to be higher than point-of-use treatment. And lastly, central treatment was considered too complicated to be efficiently operated in a community the size of San Ysidro. The results of the study indicated the best solution to be point-of-use treatment with reverse osmosis units. A pilot unit, a Culligan H-82 with a spiral-wound polyamide membrane, was installed in the community center to assure the effectiveness of the membrane and the acceptability of the unit to the community. This unit makes about 5 gpd of water with a reject rate of about 10 to 20 gpd, which is discharged into the user's septic tank. That test unit has now been in service for about three years with little maintenance required.

Since arsenic and fluoride are only considered harmful in water used for human consumption, a point-of-use unit for treatment was needed for only water used for drinking and cooking. A single large RO unit for only the drinking and cooking water supply for the village was considered, but there were still concerns about disposal of larger quantities of reject water and there was also doubt that the people would be as willing to use the treated water if they had to travel somewhere to get it. The EPA was also very interested in trying point-of-use in a small town. All of these factors made the decision to try the units in the individual homes a fairly easy one. It was decided that the best place to install the treatment units would be in the home's kitchen, preferably under the kitchen sink with a separate faucet on the sink for dispensing the treated water and a small tank under the sink for storing the treated water.

Once it was determined that point-of-use reverse osmosis might be a good solution for the village, a proposal was made to the EPA to obtain a grant to purchase, install, service, and monitor the units and to study the overall feasibility of point-of-use treatment in a small community. A Request for Proposal for engineering services was generated and Leedshill-Herkenhoff was retained by the village to oversee the project. A Request for Proposal then was written to obtain the units in addition to a maintenance contract for a period of 14 months. Culligan was awarded the job and unit installations began in June 1986. A public hearing was held in which the proposal was brought before the villagers to explain the problem with water quality and to discuss the procedures needed to get the units installed, maintained and tested during the study period. An ordinance was passed by the village which made the use of village water contingent upon installation of the RO unit in the home. Each water customer also had to sign a permission form to allow the village to install the unit in their home and to allow access to the unit for testing and maintenance. A few reluctant villagers did not want the units installed in their homes. The primary reason given was that they did not think they needed them. After all, people in the town had been drinking the water for years and it did not seem to hurt them. Another reason was the permission they had to give the village to be able to enter their homes to install, test, and maintain the units. The reluctant few were inevitably persuaded, however, when they were informed their water was going to be shut off if they did not comply. There are still a few people in the community who do not drink the water from the RO units. They say they do not like the taste of the treated water and are either getting water elsewhere or drinking the untreated water.

Currently, 70 units are available for testing on this project. There are a few units in unoccupied homes which the village has been reluctant to take the initiative to remove. The 70 units are tested every other month for arsenic and fluoride and approximately every three to four months for chloride, iron, and manganese. A smaller sample group of about 30 units is being sampled for bacteria. The testing portion of this project has been difficult at times because of the inability to obtain samples when homeowners are away. Some residents of San Ysidro are home only on weekends, and many work during the day. The sampling has been done by a village employee. He has reported that he has trouble from close to half the residents when getting samples. Most of the complaints are about the inconvenience of having to let him in. One resident draws her own samples because she refuses to let the village employee in her home despite the permission to enter that she signed. It usually takes 2-1/2 days to collect samples from 25 to 30 homes.

Because of the time frame involved, we have only been able to obtain 10 bacteria samples during each sampling. These are usually taken on a weekday morning. All of the samples are picked up and taken to an independent laboratory in Albuquerque the afternoon of the day the bacteria samples are drawn. There have been occasional problems with coliform counts in the RO units. Out of 96 tests that have been performed over the last 10 months, there have been six positive tests ranging from one to TNTC (too numerous to count). The tests have also been differentiating between coliforms and noncoliforms. We have been getting positive noncoliform readings also, almost always in conjunction with a high coliform count. Our procedure has been to have Culligan replace the filters and the RO module and disinfect the tubing and tank on the units that have shown coliforms in the treated water. The EPA lab in Cincinnati will be attempting to identify the species of coliform we are seeing as soon as we get another positive test. (This was done recently and the coliforms were identified as *E. coli.*) Culligan has recently recommended an alternative disinfection technique utilizing hydrogen peroxide that will be less expensive. The replacement costs of the filters and membrane run about $200 (see author's note).

There are a few possibilities for the source of the coliforms that we have been trying to pinpoint. It is possible that the low system pressure may be inducing back siphonage from some cross-connections in the individual homes. It was recently discerned that many of the units installed in San Ysidro had not been installed with an air gap on the discharge line from the RO module. These could very well be the cross-connections that have been causing us to see coliforms in testing. Discussions with the installer of the units revealed some disinformation regarding the air gaps. He felt the air gaps were not really necessary and that it was just one more place for the units to develop leaks. In further discussion it was explained that without the air gaps, especially in San Ysidro, where we have frequently seen low or no pressure on the water system, the likelihood of back siphonage is much greater. The installer is currently rectifying this problem.

It is also possible that the source could be somewhere in the system. The village water supply is chlorinated by a hypochlorination system at the infiltration gallery. The village has had problems in the past with the chlorination system. San Ysidro was on a boil order a few years ago for a coliform infraction prior to the installation of the RO units. During the test period we have not had a positive coliform test in the system, but the monthly system sample is obtained very close to the chlorinator. The piping system in the village is arranged in a three-spoke system with the water supply at the hub and the pipes dead-ended at the edges of town. This arrangement could encourage bacterial growth in the stagnant ends of the pipe but the positive test locations do not seem to support this theory. The locations of homes with positive results are not in any particular location on the system. The State of New Mexico Environmental Improvement Division (EID) has encouraged the village to monitor the chlorine residual and even provided them with a monitoring device, but they have been uninterested in using it. If an adequate chlorine residual was maintained in the system, it might reduce the coliform problems in the units by removing a potential bacteria source. We will be recommending to the village that they start monitoring their system chlorine residual especially at homes that show a positive bacteria count in the RO unit. The Village should also retest the treated water and test the untreated water at that home after a positive test result. In the one home in which we were able to do this, the test had shown six coliforms and three noncoliforms. The retest showed 0 coliforms and 0 noncoliforms in the RO treated water and 137 noncoliforms in the regular sink water. These retests could prevent some unnecessary maintenance costs. We will also be recommending to the village that we continue testing each unit for bacteria at least every three months until all of the air gaps are installed properly and they have the new pumping system maintaining a minimum system pressure of at least 20 psi, and until they have not had a positive test result for at least six months. After those conditions are met, we feel they should be able to decrease testing to every six months for each unit.

Initial evaluations of the test results obtained from the RO units indicate removal rates for arsenic, fluoride, chloride, iron, and manganese to be consistent with the manufacturer's data for the units despite the low system pressure experienced frequently in San Ysidro. The arsenic and fluoride testing seem to indicate that some membranes may need to be replaced as often as once a year. We had five units that had fluoride and arsenic tests at or close to the MCLs in July and August. It appears there will be two to three months' warning on these breakthroughs. The test indicator on the unit was also showing a red light on four of these units, so this will help the customer to determine if his unit needs servicing. Since the testing has shown that fluoride tends to break through just prior to the arsenic, we recommend monitoring fluoride levels in the units every three months to determine when the unit needs servicing.

The most prevalent maintenance problems are leaks. The most likely place for the leak to occur is at the drain clamp. This problem is predominant because the clamp is located where it is easily bumped and it is just tightened around the pipe rather than being attached to it. The second most common problem is breakage of the faucet handle. San Ysidro anticipates having their village maintenance man trained by Culligan so that they can take care of routine problems themselves and thus reduce their future maintenance costs.

Expenses for the study period have been broken down as follows:

Initial units	$290.00 per unit
Installation	35.50 per unit
Maintenance	8.60 per unit per month
Testing	25.00 per unit per month
Replacement or New Units	350.00 per unit

The future costs should run approximately the same except for the testing costs. It will be possible to virtually eliminate the laboratory testing for arsenic and fluoride by monitoring the conductivity of the effluent water. This is the theory behind the RO test module currently installed in each unit in San Ysidro. If the mathematical correlation between conductivity and fluoride and arsenic concentration is established, each unit can be tested on-the-spot by the village employee during a periodic inspection. Laboratory testing could be reduced to random sampling every six months to insure continued correlation between concentration and conductivity. This procedure was discussed with and tentatively accepted by our state EID representatives. The costs that we do not have information on as yet are the insurance costs for liability and replacement. However, the reduced costs of using a village employee for routine unit maintenance should at least offset the insurance costs. This would increase the user's bill by approximately $12 per month (the current bill averages $10 per month). Preliminary studies by Dennis Clifford indicated costs of point-of-use RO to be $10 to 15 per customer per month and the costs of central RO treatment to be $30 to 40 per customer per month based on the current consumption rate (see author's note).

We are currently working with San Ysidro's attorney and the State of New Mexico Environmental Improvement Division to develop an ordinance to govern the policy on the RO after the grant term is up and the village is under the state's jurisdiction again. We anticipate that the village will continue to own, maintain, and test the units in the future. The village will also have to obtain a special liability policy for the units for any water damage claims that may be made. Another insurance issue that came up during the study period was damage to the unit. We had one unit destroyed in a house fire and two units needed various major parts replaced because of freeze damage. In the future we feel the way to handle these issue will be to have the homeowner be responsible for these costs.

Special provisions will be needed for commercial establishments. We feel the best way to handle this issue in the future will be to have the village lease a properly sized unit to the business, arrange for a maintenance contract with the manufacturer, and then add these charges to the business' water bill. This should be easier than having the businesses or the village purchasing the units outright and should also give the village flexibility with new businesses that may require smaller or larger units.

On the whole, community reception to the units has been positive. Most villagers like the taste of the treated water, especially for coffee and ice. There are still some residents, primarily those who have lived in San Ysidro all their lives, who do not like and do not drink the treated water. They still drink the untreated water. There are also a couple of residents who still bring in their drinking water from elsewhere. There are a few villagers who are on the water system but who do not have indoor plumbing, a sink, or other convenient place to install the unit. They have expressed a desire to obtain the units but have had to be turned down until they have a place for the unit to be installed. There are also some residents who are dissatisfied with their individual wells who are considering getting on the village water system and are interested in the RO units. Eighty units were purchased initially for the Village and currently 79 are installed. This number has been sufficient for the study period, but a few extra will be purchased so there will be some spares available in the near future.

In conclusion, I would like to summarize the pros and cons that I've seen with this project. First the negative issues: sampling costs can be much higher for multiple point testing of point-of-use systems than single point testing of central treatment. With point-of-use, control of the treatment process is dispersed from a central point to multiple points. To maintain the same level of control, more regulating and monitoring must take place. Obtaining samples can be difficult and time consuming, contributing to higher costs and decreased control over testing. With a point-of-use treatment system, another factor that must be considered is initial and continuing education of the consumer. New members of the community must be indoctrinated to the system, and existing users must be reminded periodically of their responsibilities. Point-of-use also inherently generates more bookkeeping for the village clerk. The responsibility for tracking, testing, and maintenance for each unit will be a part of the job. Another problem that is an inherent part of point-of-use treatment is do-ityourself plumbers. It is an irresistible urge for some people to tamper with or try to fix the units themselves since the unit is located in an accessible part of the home.

The positive side of point-of-use in San Ysidro is much more encouraging. The RO units have very much improved the aesthetic as well as health quality of the village's drinking water. The units are simple to install and maintain. The units are undoubtedly the least expensive solution to treatment for the village and its residents. The vast majority (90 to 95 percent) are happy with the system and the water it produces.

This project has been challenging for the community of San Ysidro. If the units are to continue to function as part of the village's treatment system, the community will need encouragement and technical and regulatory support from the State of New Mexico's Environmental Improvement Division. The village will need to begin recordkeeping on each unit and diligently maintain those records to insure the maintenance and testing continues as required. We will also have to see how the community will respond to the additional costs on their water bills. The future success of this project will depend heavily on the abilities of the Village of San Ysidro to cope with the recordkeeping for testing and maintenance of the units, but this would surely be a problem for central treatment as well. Point-of-use treatment is the best choice for San Ysidro's water system at this time and will continue to be the best solution until population growth in the village will support the costs of a central treatment plant.

Author's Note: This report has been revised to reflect pertinent additional data and information obtained since it was presented.

FLORIDA'S FUNDING FOR CONTAMINATION CORRECTION

Glenn Dykes
State of Florida
Department of Environmental Regulation
Tallahassee, FL 32399

For years Florida has had considerable concern for drinking water quality because of the expanding use of its ground water resources. The potential for contaminating these resources is great in view of the state's rapid growth and the vulnerability of this valuable source of potable water. Water quality problems have been confirmed by extensive analytical work, and now these contamination issues must be addressed.

DEFINING THE PROBLEM

Over 90 percent of the state's drinking water comes from underground resources. Included in these supplies are approximately 10,000 wells serving public and semi-public facilities along with several hundred thousand private wells, serving individual homes. There is a wide disparity in the quality and the vulnerability of the ground water resources meeting the demands of these water supply wells. Many of the public systems utilize the deeper limestone strata while the private wells primarily tap the shallower aquifers. These two sources have different problems and concerns, but both are susceptible to contamination that has been verified by analytical work on these resources.

Early in the organic quality testing work by the U.S. EPA, several of Florida's public supplies were analyzed. The EPA efforts and testing by the state's own laboratories confirmed suspicions of contamination of the state's ground water supplies. From these early endeavors, the state developed a very comprehensive set of rules and regulations governing ground water protection, as well as the first regulations in the nation setting maximum contaminant levels (MCLs) for volatile organic compounds. The driving force for these actions was the realization that the agricultural chemicals, aldicarb and ethylene dibromide (EDB), had also been found in samples from many of the private wells. From further assessment of this problem, it was also learned that these chemicals, particularly EDB, had widespread usage throughout the state. Consequently, full evaluation and correction of all possible contamination would take considerable money, time, and effort. An important aspect of the use of EDB was the fact that the state, under contract to the citrus grove owners, applied the chemical for nematode control. Since the state and Federal agriculture departments approved of the chemical usage, it found broad acceptance in the agricultural community.

LEGISLATIVE EFFORTS

The findings of organic contamination in both the public and private potable water supplies brought about vocal manifestations of the electorate's concerns. The legislature started looking for ways to correct these problems. In 1983 they passed the Water Quality Assurance Act (WQAA), which was very broad legislation addressing a wide variety of items related to ground water contamination and providing funding for their resolution. This act spelled out numerous ground water protection issues, which included requiring public water suppliers to investigate a broad spectrum of contaminants, studying private wells, and funding related research and emergency corrective actions. Law makers the following year provided $3.1 million for solving the EDB contamination problem caused by the state's own activities. The 1986 legislature broadened the WQAA and provided additional funding to correct EDB contamination and other health related water quality problems over which the well owners had no control. This legislature also provided funding to address contamination from leaking underground petroleum storage tanks through the establishment of the Inland Protection Trust Fund (IPTF). Backed by this authority, the Florida Department of Environmental Regulation (FDER) could restore or replace contaminated private wells (WQAA and IPTF) and public wells (IPTF) without waiting for legal determinations of responsibility. Funds for correcting problems in public supplies were not available under WQAA though assistance was provided with the state-affected EDB monies.

RESTORATION/REPLACEMENT OF SUPPLIES

The Ground Water Contamination Task Force, which was organized to address EDB and other contaminant problems, had to evaluate the potential concerns and how to approach each one. Initially, activities were directed at EDB. The group determined that correction would have to be provided for the entire household, mandating point-of-entry treatment as the only approach. Through research funding, data were developed to indicate that granular activated carbon (GAC) would effectively adsorb EDB. The state collaborated with EPA to fund a project to provide definitive numbers for both packed tower aeration and GAC that would assist in treatment unit designs. Research funds were also used to evaluate the viability of replacing contaminated supplies with properly constructed wells into the deeper aquifers. In the final analysis, this approach did not prove feasible for widespread application.

The use of GAC filters was determined to be the best alternative to correct the numerous private wells contaminated with EDB. Since the volume of water, usage habits, and other parameters that influence the effectiveness of GAC were all undefined, a very conservative filter design was devised. A 0.06-m^3 (2-cu ft) GAC filter was selected with a 5-μm (0.0002-in) pre-filter, a water meter, and an ultraviolet (UV) light included in the standard unit (Type I). These units were installed to handle contamination up to 10 μg/l. Higher contamination warranted additional GAC filter units (Type II). If the expected water consumption was more than 10 gpm, larger GAC units were provided to handle increased flow. Formal bid proposals were solicited to obtain a qualified supplier, since it was realized that a large number of units would be needed. The proposals also included operation and maintenance items such as the planned replacement every six months of GAC material and UV lights.

Of the 12,400 wells analyzed for EDB through October 1987, 1,530 were found to exceed the MCL of 0.02 μg/l. Most of these (~1,400) served private residences. Where possible, a permanent solution to correct the problem, such as connections to existing community supplies, was utilized. At present we have over 550 Type I and 60 Type II units installed on state-affected wells. There are also 230 Type I and 2 Type II GAC filters on non-state-affected wells. Because of different statutory requirements, records on each program are separately maintained. The overall EDB corrective effort has also required the installation of larger units on some of the public and semi-public systems. There are seven larger GAC units between 50 and 200 gpm capacity and three municipal systems with capacities up to 3,000 gpm.

Under the state's current contract, the Type I installation cost is $1,000, and the cost of Type II units is $1,050. The annual carbon and UV light replacement cost is $890 for the residential type unit. The overall program is currently being re-evaluated to determine if the longevity of the GAC filters can be extended past the current six-month replacement cycle. It is estimated that each month's extension would save approximately $40,000. If the foregoing GAC units and maintenance costs are evaluated, one can easily see that annual outlay is quite large. The annual maintenance cost for the residential units and the aforementioned larger systems will exceed $1 million when the remaining planned units are installed.

In view of the potential future maintenance cost, the state has attempted to evaluate the economic feasibility of extending existing water lines to replace the residential supply wells. To determine the current cost of the filter installations for their lifetime, the annual replacement cost was projected for 10 years and then returned to present worth using five percent inflation and eight percent interest. This value was then added to the installation cost of the filter to determine the most economical approach; i.e., individual GAC filters, extension of water lines, or a new central system. Under the current contract, the calculated cost for use in this feasibility analysis is approximately $8,600. Since additional organic sampling is required to insure that public health concerns are not compromised, an additional annual cost of $400 should be added. This would add approximately $3,400 to the aforementioned present cost consideration. With the logical addition of sampling cost, $12,000 should be used to determine the economic feasibility to provide a permanent alternative to the point-of-entry solution.

SUMMARY AND CONCLUSIONS

Carbon filtration has been found to be a satisfactory method of removing EDB from contaminated supplies. The FDER is also utilizing GAC filters in correcting contamination problems created by leaking underground petroleum storage tanks under the IPTF program. Current testing indicates that GAC will remove benzene and other hydrocarbons and thereby solve some of the problems related to these contaminants.

The use of point-of-entry solutions must consider the long term cost in considering the economic viability. The cost as shown in the foregoing discussion can be large. The annual maintenance and sampling cost must be given a thorough evaluation before determining that point-of-entry devices are to be placed on all residences in a community. It is always gratifying when a permanent solution can be found and the utility does not have to worry about the maintenance problems that always seem to plague these small installations. The present projected cost of $12,000 per house with a Type I connection would go a long way to provide a central system for the whole community.

MONITORING AND MAINTENANCE PROGRAMS FOR POU/POE

Gordon E. Bellen
Thomas G. Stevens
National Sanitation Foundation
Ann Arbor, MI 48105

INTRODUCTION

Small communities with organic or inorganic contaminants in their drinking water supplies often lack the financial resources to solve their problems. Economies of scale prohibit construction of a central treatment system for contaminant removal in many cases. Construction of an alternate well or connection to a neighboring water supply may not be feasible. One alternative solution, which has been receiving more attention in recent years, is treatment of contaminated water at the point-of-use (POU) or point-of-entry (POE).

All POU devices are designed to treat only water intended for consumption. Approaches to POU treatment include batch process treatment, faucet-mounted devices, in-line devices, and line-bypass devices. A batch process device treats one batch of water at a time, is not connected to the water supply, and may rest on the kitchen countertop. Faucet-mounted devices are attached directly to the faucet. In-line devices are installed between the cold water supply and the kitchen faucet, and generally treat the entire kitchen cold water supply. With the line-bypass approach, the cold water line is tapped to provide influent to a treatment device, which may be installed under the kitchen sink, and a separate tap is provided at the sink for treated water.

POE water treatment treats all water entering the home and has been proposed for contaminant removal where potential health risks associated with skin contact and inhalation exist (1). Because they treat all water entering the home, POE devices must be much larger (in terms of volume treated) than POU devices. The length of time in service between media replacements however, is typically 25 percent of that of POU devices.

The U.S. EPA has specified when, and under what conditions, POE and POU water treatment can be used (2). Although not considered Best Available Technology (BAT), POE is an acceptable method for a community water supply to come into compliance with the Drinking Water Regulations. POU may be used as an additional control measure during the period of a variance or exemption, as a condition of the variance or exemption. If either approach is used, the EPA has specified conditions that must be met:

- *Central Control* - Regardless of ownership of the treatment device, the public water authority will be responsible for operating and maintaining all parts of the treatment system.
- *Effective Monitoring* - A monitoring plan must be approved by the state before a POU or POE system is installed. The plan must assure that devices provide health protection equivalent to central water treatment. Physical condition of equipment and total volume of water treated must be monitored as well.
- *Application of Effective Technology* - All devices must have certified performance (or rigorous design review) and must pass field testing.
- *Maintenance of Microbiological Safety* - Control techniques such as backwashing, disinfection, and monitoring, are suggested by the EPA to maintain microbiological safety.
- *Protection of All Consumers* - Every building must have equipment that is adequately installed, monitored, and maintained. Responsibilities for this equipment may transfer with ownership of the property.

This paper discusses each of these points and suggests ways in which a community might comply with EPA requirements.

CENTRAL CONTROL

It is important that all aspects of POU/POE treatment come under central control to assure adequate protection of public health. If a public water system is already in place, the existing organization can assume administration of the POU/POE district. If a public water system is not in place, a water quality district should be formed in a progression of steps similar to those in Table 1 (3). In a previous study (4), total administrative costs for operating a water quality district were estimated (1985 dollars) to be $1.23 per customer per month. These costs included quarterly monitoring costs, administration, and distribution system maintenance (POU maintenance not included) for reduction of fluoride in drinking water with POU treatment. Monitoring costs may be higher for some contaminants, but labor costs can be lower if community volunteers are used.

Table 1. Chronological Steps for Formation of a Water Quality District

Process Step
Development phase
• Identify problem
• Consult regulatory agencies
• Water testing
• Make preliminary plans and maps
Approval phase
• Estimate costs
• Hold public hearing
• Property owner petition
• District formed by resolution of county/state supervisors
• Directors appointed
• Agreement with town board and property owners for cost recovery
Operation Phase
• Obtain funding
• Pilot demonstration
• Select equipment
• Equipment installation
• Authorize payments
• Monitoring and maintenance
• Feedback and education

APPLICATION OF APPROPRIATE TECHNOLOGY

To achieve compliance, the EPA stipulates application of appropriate technology. Table 2 lists currently available technologies and the contaminants they are effective in removing. Selection of appropriate technology from this list is not straightforward, since variable water qualities may make one technology better than another. For example, activated alumina is effective in fluoride reduction, but in the presence of high alkalinity and/or arsenic, its capacity may be reduced (4,5).

Communities lacking expertise should seek knowledgeable sources of information concerning treatment techniques appropriate for their water. An initial consultation with the local or state health department is a good first step. Other organizations that can provide information are listed in Table 3. In addition to those organizations, consulting engineering firms can be hired. Regardless of the source used, professional guidance in the selection of equipment is important.

Table 2. POU/POE Treatment Technologies*

Treatment Type	NIPDWR Contaminants	Other Contaminants
Reverse Osmosis**	Arsenic***, Barium,, Cadmium, Chromium, Lead, Mercury, Silver, Fluoride, Nitrate, Selenium, Radium, some organics, herbicides, and pesticides	Total dissolved solids, Copper, Chloride, Sulfate,foaming agents, corrosion
Cation Exchange	Barium, Cadmium, Chromium III, Lead, Mercury, Radium	Copper, Zinc, Iron†, Manganese
Anion Exchange	Nitrate, Selenium VI, Arsenic III, Arsenic V, Chromium VI	Chloride, corrosion, Sulfate
Activated Alumina	Fluoride, Arsenic,Selenium IV	
Direct (Mechanical) Filtration	Turbidity	Cysts
Activated Carbon	Organics, Organic Mercury	Color, foaming agents, taste, and odor
Distillation	Metals, high molecular weight organics	Total dissolved solids, Chloride, Sulfate

* Taken from the *Statement of the Water Quality Association to the EPA*, EPA, December 13, 1983.
** Results of reverse osmosis treatment may vary between pressurized and nonpressurized units, membrane type, and configuration.
*** Arsenic ($^{+3}$) is poorly removed with reverse osmosis.
† Low levels.

Table 3. Organizations Providing Water Treatment Information or Services

Organization	Service
American Water Works Association 666 West Quincy Avenue Denver, CO 80235 303/794-7711	Water Data Base Educational Materials Technical Information
National Demonstration Water Project 1725 DeSales Street, NW - Suite 402 Washington, DC 20036 202/659-0661	Educational Information Relating to Rural Communities
National Sanitation Foundation 3475 Plymouth Road Ann Arbor, MI 48105 313/769-8010	Product Testing/Listing Performance Standards Technical Information Technical Assistance
National Water Well Assocation 500 West Wilson Bridge Road Worthington, OH 43085 614/846-9355	Ground Water Information
Water Quality Association 4151 Naperville Road Lisle, IL 60532 312/369-1600	Lists of Manufacturers and Distributors Technical Information

The EPA requires certification of performance and field testing of POU and POE devices. Certification can be accomplished by the state or a third-party acceptable to the state (2). The National Sanitation Foundation (NSF) has several performance standards for POU/POE devices that address performance for products making contaminant reduction claims for primary and secondary regulated chemicals, reverse osmosis equipment, cation exchange water softeners, and ultraviolet disinfection equipment. A standard for distillation equipment is currently being written. These standards are in the public domain and can be used by anyone as a basis for product certification. NSF also conducts a product certification program. NSF listings of products are available on hard copy or through computer access (see Table 3 for address and phone number of NSF).

Field testing is important. Product certification testing with standardized test waters may not accurately indicate treatment capability or capacity for the water a community needs to treat. For treatment technologies with finite usable capacities (e.g., carbon and activated alumina), field tests should be run to exhaustion to establish the useful life of the device (useful life defined as volume of water treated to breakthrough). Accelerated field tests using surface adsorbents like carbon and activated alumina will provide conservative estimates of useful life. Tests of devices with media which can be regenerated will provide estimates of annual regeneration costs. Other devices like reverse osmosis and distillation, that do not have a readily definable useful life, can be evaluated quickly for percent removal of undesirable contaminants.

EFFECTIVE MONITORING

The EPA requires monitoring to assure protection of the public health comparable to central treatment. To achieve that goal, a monitoring program should be established which provides reasonable assurance that all water provided at the tap is in compliance with the National Primary Drinking Water Regulation. The results of field testing should help in that regard. In addition, the total volume of water treated and physical condition of each unit must be monitored (2).

The system performance monitoring program will be influenced by field test results, community experience, and whether treatment is intended to achieve compliance. A rigorous and conservative monitoring program, assuming one technology or type of device is being used, is outlined in Table 4. This program addresses POU/POE treatment on a distribution system. Treatment of a system of individual wells would require more frequent sampling. This monitoring program is intended to assure adequate operation of the POU/POE treatment system. Additional source water monitoring is required by the EPA. A description follows.

Table 4. Suggested Minimum Monitoring Program*

Task	Frequency
Contaminant Monitoring	
First Year	Minimum of seven devices per quarter; if useful life is less than one year, test at each quarter of estimated life.
Second Year	Minimum of seven devices per quarter; select some known high volume users.
After Useful Life Established	Minimum of three per quarter. Test minimum of seven units at replacement to reconfirm useful life estimtes.
Microbiological Sampling	
Routine Monitoring	
Heterotrophic Plate Counts and Coliforms	Minimum of seven per quarter or number required by EPA population based monitoring plans, whichever is higher.
Fecal Coliforms	If positive coliform results obtained.
Treated Volume Recordings**	(May be provided by homeowners)
First Year	25 percent of devices/quarter.
After First Year	10 percent of devices/quarter.

* In addition to other system monitoring requirements.
** Volume recorder for treatment unit, not whole house meter.

Field test results should be used to estimate useful life of component parts (e.g., media, cartridges, prefilters, etc.), although manufacturers and/or consultants can also provide guidance. In addition, a sampling plan that confirms total community compliance is necessary. The plan should provide the minimum number of samples to accurately and statistically represent the number of installations. In this example, seven installations were assumed to be the sample size. If the useful life is one year or greater, sample quarterly. For less than one year, sample at intervals of 25 percent of useful life. At least seven of the devices should have effluent samples checked immediately after installation. Continue quarterly sampling of seven of the devices so that devices have been tested for treatment performance throughout the first year of operation. Quarterly sampling of a minimum of seven devices should be continued until the useful life of all components under normal operating conditions can be more precisely defined. Once this more precise useful life has been established, replacement can be tracked based on the volume of water treated. Actual performance sampling can then be based on recommended sampling frequencies for central water supplies.

If an effluent sample from a device is positive prior to the estimated useful life, resample to confirm that breakthrough has occurred. Replace the treatment component of a device if breakthrough is verified. Test an additional seven devices to determine if early breakthrough is occurring throughout the system.

The sampling may be reduced as community experience increases. In most cases, it will be less expensive to replace components prematurely than to sample frequently enough to optimize component life. Water meters are available with automatic shut-offs, alarms, and even telemetry. A comprehensive treated water volume monitoring system can be easy and economical to establish and operate. User histories should be established to guide meter reading schedules.

If useful life is monitored based on gallons of water treated, one of the more important effluent samples for the device becomes the sample taken immediately after installation. This sample serves two purposes: 1) it assures that the device is operating, and 2) it can identify media contamination. Media contamination is rare, but has been noted on occasion (6,7).

The physical condition of the device should be verified upon installation and an operational check of the device should be part of the installation procedure. Most problems with faulty devices or installations will occur within the first few weeks of operation (4,6,7). After this initial period, spot checks of installations coincident with monitoring and/or meter readings should suffice. However, homeowners should have access to 24-hour repair service.

MAINTENANCE OF MICROBIOLOGICAL SAFETY

The EPA requires that communities using POU/POE devices for compliance treatment assure that the treated water is microbiologically safe. However, microbiological safety has not been clearly defined. Testing for coliform organisms does not necessarily indicate the presence or absence of other pathogens. Heterotropic plate counts are even more ambiguous.

Heterotrophic bacteria will colonize on carbon and other surfaces, but efforts to colonize pathogens on carbon in the presence of competing bacteria have not been successful (8,9). The infrequent contact of pathogens in a water supply with a POU/POE device should not result in a colonization of pathogens (4,6,7). Therefore, POU/POE devices may pose no greater risk of increased pathogens than if they were not installed. POU/POE devices should not be used with water of unknown microbiological quality (10,11).

A monitoring program should include heterotrophic plate counts and coliform counts. The EPA has determined that it is important to keep heterotrophic plate counts below 500 per ml to reduce interference with coliform counts (12). It has been demonstrated that flushing (running water through) devices will reduce heterotrophic counts (4,7,10). Consequently, it is important to use standard water sampling methods for microbiological analyses. These methods include disinfecting the sample tap and running water for two minutes prior to sampling (13), which should provide adequate flushing.

If devices show greater than 500 organisms per ml, using this procedure, the device or component should be replaced. In place disinfection of a device or component is not recommended since bacteria colonized on carbon are less susceptible to disinfection (9).

During the first year of operation, monitoring should include sampling a minimum of seven devices per quarter for microbial analyses. More sampling will be necessary for larger communities. Standard EPA community sampling frequencies for coliforms based on population served should be followed (12).

In addition to monitoring, preventive measures can also be taken. Silver-impregnated devices may provide some protection against coliform organisms, but they will not typically reduce hererotrophic plate counts (7,10,14). The addition of a POU/POE disinfection device is also an alternative. Table 5 lists disinfection technology that can be applied with POU/POE devices (15).

Table 5. Currently Available Water Disinfection Technology Applicable to POE/POU Treatment

Chlorination
Liquid Chemical Feeders
Other Halogens
Resin Based Brominators
Resin Based Iodinators
Ozonators
Electrolytic Generation
Ultraviolet Light
Flow Through Irradiation

SUMMARY

Providing water to consumers that meets the U.S. EPA National Primary Drinking Water Regulations requires system organization, maintenance, and monitoring. This is true whether central, POU, or POE treatment is used. The goals for protection of health are the same regardless of the method used to attain the goals.

While POU and POE technologies are not recognized as best available technologies in the regulations, they are considered to be acceptable for use if specified conditions of system control, monitoring, effectiveness, and public health protection are assured. Effective and reasonable monitoring and maintenance programs can be developed to meet the requirements of the regulations, whether within an existing water system organization or by means of an organization established specifically for operation and maintenance of POU/POE systems.

For water treatment professionals with experience only in central treatment, it may, at first, seem extremely difficult to achieve comparable health protection with POU or POE devices. However, as experience is gained with new approaches to community treatment, POU or POE water treatment districts may offer attractive benefits for some communities.

REFERENCES

1. Andelman, J. Non-ingestion exposures to chemicals in potable water. Center for Environmental Epidemiology, University of Pittsburgh, Pittsburgh, PA, 1984.

2. National primary drinking water regulations; synthetic organic chemicals; monitoring and unregulated contaminants. Federal Register Vol. 52, No. 130, July 8, 1987.

3. Bellen, G.E., Anderson, M.A and Gottler, R. Management of point-of-use, drinking water treatment systems. Final Report U.S. EPA Contract R809248010. National Sanitation Foundation, Ann Arbor, MI, 1985.

4. Bellen, G.E., Anderson, M.A and Gottler, R. Defluoridation of Drinking Water in Small Communities. U.S. EPA Contract No. R809248010. National Sanitation Foundation, Ann Arbor, MI, 1985.

5. Singh, G. and Clifford, D.A. The equilibrium fluoride capacity of activated alumina. Project summary, EPA-600/52-81-082. July 1981.

6. DeFilippi, J.A. and Baier, J.H. Point-of-use and point-of-entry treatment on Long Island. Journal AWWA, Vol. 79, No. 10. October 1987.

7. Bellen, G.E., Anderson, M.A. and Gottler, R. Point-of-use reduction of volatile halogenated organics. U.S. EPA Contract R809248010. National Sanitation Foundation, Ann Arbor, MI, July 1985.

8. Geldrich, E.E. et al. Bacterial Colonizing of point-of-use water treatment devices. Journal AWWA, Vol. 77, No. 2. February 1985.

9. McFeters, G.A. et al. Bacteria attached to granular activated carbon in drinking water. U.S. EPA 1600/M-87/003. Cincinnati, OH.

10. Reasoner, D.J. et al. Microbiological characteristics of third faucet point-of-use devices. Journal AWWA, Vol. 79, No. 10. October 1987.

11. National Sanitation Foundation. Standard 53: drinking water treatment units - health effects. Ann Arbor, MI (revised June 1982).

12. National primary drinking water regulations; filtration and disinfection: turbidity, *Giardia lamblia*, viruses, *Legionella*, and heterotrophic bacteria; proposed rule. Federal Register Vol. 52, No. 212. November 3, 1987.

13. Standard methods for the examination of water and wastewater. 16th edition, APHA. Washington, DC, 1985.

14. Regunathan, P. and Bauman, W.H. Microbiological characteristics of point-of-use precoat carbon filters. Journal AWWA, Vol. 79, No. 10. October 1987.

15. Bellen, G.E., Gottler, R.A. and Dormand-Herrera, R. Survey and evaluation of currently available disinfection technology suitable for passenger cruise vessel use. Centers for disease control. Contract No. 200-80-0535. National Sanitation Foundation, Ann Arbor, MI. September 1981.

POINT-OF-USE AND POINT-OF-ENTRY TREATMENT DEVICES USED AT SUPERFUND SITES TO REMEDIATE CONTAMINATED DRINKING WATER

Sheri L. Bianchin
U.S. EPA-Region V
Chicago, IL 60604

Hazardous waste is one of this nation's greatest concerns. In response to that concern, a law was enacted to deal with the hazardous waste problem. This law is the Comprehensive Environmental Response, Compensation, and Liability Act (CERCLA), and is referred to as the Superfund Law. This law provided broad Federal authority and resources to investigate and to respond directly to releases (or threatened releases) of hazardous substances that may endanger human health or the environment. Costs for the first five years of the Superfund program were covered by a $1.6 billion Hazardous Substance Response Trust Fund established to pay for cleanup of abandoned or uncontrolled hazardous waste sites. The law also authorized enforcement action and cost recovery from those responsible for the release.

CERCLA was revised in 1986 as the Superfund Amendments and Reauthorization Act (SARA). The purpose of the revision was to renew and strengthen the Superfund Program. SARA reauthorizes the program for five years and increases the size of the fund to $8.5 billion.

SARA gives the United States Environmental Protection Agency (U.S. EPA) the authority and responsibility to control the actual or potential release of hazardous substances that pose a threat to human health or welfare or the environment. Other Federal agencies will provide assistance as necessary during response. A comprehensive regulation known as the National Contingency Plan (NCP) describes the guidelines and procedures for implementing this law. The law, SARA, requires that hazardous waste cleanups do the following:

- Protect human health and the environment;
- Provide for a cost-effective solution with an emphasis on treatment and permanent destruction over off-site disposal; and
- Compliance with all Applicable or Relevant and Appropriate Requirements (ARAR).

ARARs are those standards or criteria promulgated under state or Federal law to specifically address the abatement of contamination by a hazardous substance, cleanup standards, or advisories.

Federal ARARs may be derived from the following:

- Safe Drinking Water Act (SDWA);
- Resource Conservation and Recovery Act (RCRA);
- Clean Water Act;
- Clean Air Act;
- Toxic Substances Control Act;
- Federal Insecticide, Fungicide, and Rodenticide Act; and
- Great Lakes Water Quality Act.

Over two-thirds of the Superfund actions to date, deal with a contaminated drinking water supply. Where SDWA Standards are applicable to the Superfund cleanup, maximum contaminant levels (MCLs) are usually used. A MCL is an enforceable standard for each contaminant, which the act directs U.S. EPA to set as close to the maximum contaminant level goal (MCLG) as feasible. Decision on the level of a MCL that is "feasible" includes consideration of the best technology treatment techniques and laboratory analyses that are available, taking cost into consideration.

On the other hand, a MCLG is a nonenforceable health goal. It is a numerical limit set for each contaminant at the level at which no adverse health effects on persons can be expected, with an adequate margin of safety.

MCLGs may be used as cleanup criteria on a site-specific determination. One factor in this determination is whether multiple contaminants or multiple pathways of exposure exist on the site.

Also important in determining the ARARs from the SDWA is the use or potential use of the water that is or is likely to may become contaminated.

The NCP lays out three types of responses for incidents involving hazardous waste. These responses are immediate removal, planned removal, and remedial.

A removal action is designed to be a short-term action to stabilize or clean up a hazardous site that poses an immediate threat to human health or the environment. Typical removal actions include removing tanks or drums of hazardous substances on the surface, installing fencing or other security measures, and providing a temporary alternate source of drinking water. Removals may be divided into two categories: immediate and planned removal. U.S. EPA's policy has been that in order for U.S. EPA to initiate a removal action for contaminated drinking water, the level or concentration of the contaminant typically should exceed the 10-day health advisory. This policy is expected to become more stringent in the near future.

An immediate removal or a time-critical removal is a prompt response taken to prevent immediate and significant harm to human life or the environment. By statute, the action must be completed within one year and the cost of the action shall not exceed $2 million. Immediate removals are taken to bring a release of hazardous substances under control; they are not intended to eliminate completely every long-term problem. Immediate implementability is the major consideration in choosing a remedy.

The other type of removal is termed a planned or non time-critical removal. This type of removal is an expedited, but not necessarily immediate, response. A planned removal action is also limited by time and monies involved in the cleanup.

Typically removal actions are completed by U.S. EPA, whereby U.S. EPA will subsequently attempt cost recovery from any identified responsible parties.

A remedial response entails a long and complicated process aimed at identifying and completing a permanent remedy to remediate and abate the hazards at a site. A remedial action is designed such that a thorough study is completed prior to the design and construction of a selected remedy. Technical measures can be selected only after evaluation of all feasible alternatives on the basis of economic, engineering, and environmental factors. Specifically addressed in a study are the ability to protect public health; technical feasibility; environmental effectiveness; ability to meet ARARs; compatibility with other Federal, state, and local laws; constructibility; reliability; cost; and community acceptance. The intent is to derive the maximum benefit from Superfund as a whole. EPA can only conduct remedial responses to those sites on the National Priority List (NPL). The NPL is a list of the nation's most serious hazardous waste sites. Typically sites are identified for listing by the state. A preliminary assessment is performed on each site. The sites are scored by the Hazard Ranking System (HRS). The HRS looks at potential pollutant pathways that may reach a receptor, like the ground and surface water pathways that may affect drinking water. Scores greater than 28.5 are listed on the NPL.

After a site is included on the NPL, a remedial action is planned in a series of defined steps. These steps are as follows:

- Remedial Investigation/Feasibility Study (RI/FS);
- Remedy Selection; and
- Remedial Design/Remedial Action (RD/RA).

A RI/FS is utilized to examine the type and extent of the contamination, and identifies and screens the possible remedies. Remedies selected must strive to be of a permanent nature. When the final decision on a remedial action or an operable unit of a remedy is reached in the remedial process, they are documented in a Record of Decision (ROD). The last phase in the remedial process, the RD/RA, is the design and construction of the selected remedy.

A Superfund action is typically funded by two mechanisms. The first type is a Superfund-funded remediation where no responsible parties have been identified, or no legal agreement for the responsible parties to conduct the work can be reached. The second type is a Responsible Party-funded remediation, where the responsible party pays for the cleanup, and EPA serves to oversee the action.

The first step in any Superfund action, whether it is a removal or remedial action, is to identify and confirm the extent and types of contamination that exist. Next the levels identified are compared with the standards and health effects information. Last, the method of correction and time required to finish the project are determined. The remedy selected will depend upon whether the action is a removal or a remedial action.

As a short-term response to alleviate the immediate danger of contaminated drinking water, bottled water is routinely utilized. This type of response is most often used in a removal action.

In the interim, the following are commonly considered alternatives where drinking water is a concern:

- Continue providing bottled water;
- Provide alternative water;
- Installation of Point-of-Use (POU) Treatment Devices; and
- Installation of Point-of-Entry (POE) Treatment Devices.

These types of responses may be used in either a removal or remedial action.

A POU treatment device is one used at the tap to purge the water of contamination prior to drinking. A

POE treatment device is one used as water enters a house to purge the water of contamination. A POE or "whole house" treatment unit treats the entire household water supply. POU/POE devices typically involve aeration (air stripping), adsorption (granular activated carbon), or reverse osmosis.

As a longer term response, or as a permanent remedy, the following are commonly considered remedial alternatives for a contaminated drinking water supply:

- Connect to a community water system;
- Provide a new water source (well or surface); and
- Maintain individual treatment unit previously installed.

These types of responses may be used in either a removal or remedial response. Next discussed will be case studies at two Superfund sites. The first case that will be discussed is the Byron Johnson Salvage yard in Byron, IL. See Figures 1 and 2 for maps of the Superfund site investigation area.

Byron Johnson is a 20-acre salvage yard located in a rural area of northern Illinois. The site is owned by three individuals, and it has been determined that domestic waste and metallic debris were deposited at the site. It is also suspected that open dumping has occurred. A description of the site history follows.

In the 1960s, the salvage yard was operated as a junk yard, where miscellaneous wastes and debris were brought for disposal. In 1970 to 1972, the Illinois Environmental Protection Agency (IEPA) conducted periodic inspections to identify any operating deficiencies. In 1972, IEPA ordered closure of the salvage yard. The salvage yard ceased operation in 1974. In December 1982, the site was placed onto the NPL. In May 1983, under agreement with U.S. EPA, the IEPA performed a state-lead RI/FS. This study specifically focused on contamination directly on or below the site; ground water contamination potentially emanating from the site was not addressed.

Through 1984 and 1985, the U.S. EPA, IEPA, and the Illinois Department of Public Health (IDPH) continued to monitor the contamination levels in residential wells located nearby and down-gradient from the site. Through periodic sampling, off-site ground water contamination by volatile organic compounds (VOCs) was documented. It was found that private wells contained trichloroethylene (TCE) in concentrations up to 710 µg/l. In June 1984, the IEPA completed a RI/FS and signed a ROD to remove drums of waste and contaminated soil from the site. In July 1984, the U.S. EPA placed the residents whose water exceeded 200 µg/l in the Dirk Farm area (i.e., those residents along Acorn and Razorville Roads) on bottled water as a temporary measure. Late in 1984, the U.S. EPA contracted to have a RI/FS performed at the site. In July 1985, a U.S. EPA action was started to augment the data collected from the IEPA RI/FS.

In October 1985, U.S. EPA conducted a phased FS to expand the scope of the study to the Rock River Terrace subdivision which is about one and one-half miles down gradient from the site. The objective of the study was to investigate the potential health threat due to the exposure to the contaminated water supply and evaluation of alternative water supply and treatment options that would ensure a safe water supply to Rock River Terrace residents. Sampling results indicated that the ground water was contaminated with levels of TCE up to 48 µg/l TCE. Although the level detected was below the 10-day health advisory, it is above the drinking water standard of 5 µg/l. Also, since the residents are located in the direction of the contaminant plume, it was determined that a planned removal action was warranted.

In May 1986, U.S. EPA, by a removal action, installed carbon adsorption POU treatment devices for those residents on bottled water in the Dirk Farm area as an interim measure to remove TCE from the water.

In June 1986, U.S. EPA completed a study that focused on the potential ground water problem for the Rock River Terrace. It was determined that both of the major aquifers in the area were contaminated to some extent by VOCs. This contamination extends to outlying locations 0.8 km (1 mi) northwest and 0.8 km (1 mi) north of the site. In addition, slight contamination by cyanide and some inorganic compounds exist in the groundwater beneath the salvage yard. In addition, the study identified and evaluated alternatives for replacing or treating contaminated water from private wells. Based upon the RI/FS, the alternatives for treating or replacing water from Rock River Terrace wells were narrowed down to the three alternatives listed below, and a detailed analysis was conducted on each one.

Alternative 1
Connection to the Byron Municipal Facility. This alternative was estimated to take one to two years to complete, and was estimated to cost approximately $900,000.

Alternative 2
Supply bottled water to homes with contaminated wells. This alternative would not provide water for bathing and washing. The annual cost of this alternative was estimated at $91,150.

Alternative 3
Treatment of water from affected wells to remove contaminants through carbon adsorption. It was estimated that it would cost $26,000 to install POU treatment devices and $115,000 to install "whole

Figure 1. Location of Byron Johnson salvage yard, Byron, IL. investigation area.

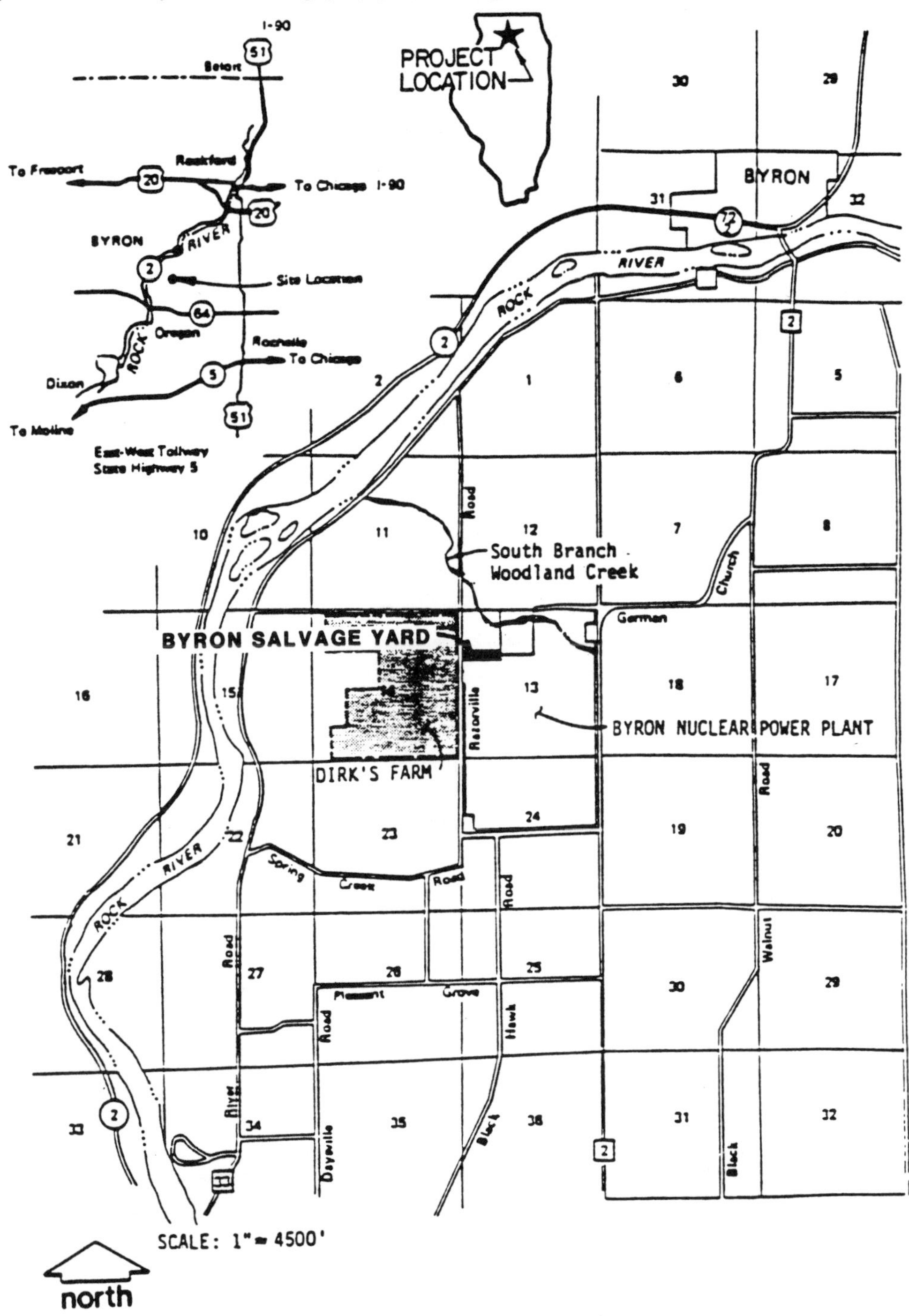

Figure 2. Superfund site investigation area, Byron, IL.

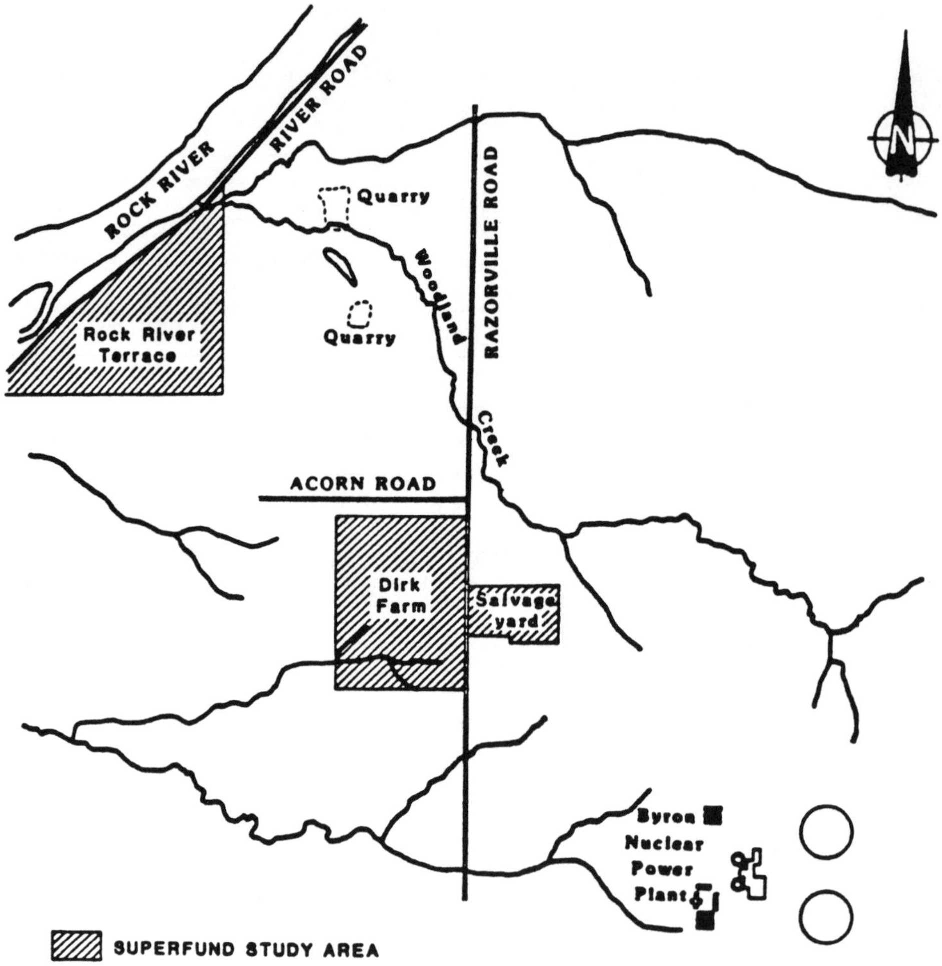

house" POE treatment devices. Upon commencement of construction, it would take approximately two to three months to install the units.

The results of the study indicated Alternative 3 to be the most economically feasible while providing a safe and reliable drinking water supply for affected residents. The whole house units would be installed at Rock River Terrace homes that are occupied on a year-round basis. Periodic monitoring would be conducted to ensure that contaminants are being effectively removed. The carbon for these units would be replaced when necessary. Since the carbon replacement rate is dependent on many factors including the level of contamination, water temperature, pH, chemical makeup of the water, and water usage, monitoring of the carbon bed is necessary.

In July 1986, U.S. EPA initiated a monthly sampling program for the residents in the Dirk Farm area to monitor the efficiency of the POU treatment devices. The units will be replaced when necessary.

In July 1986, IEPA signed a ROD for design and construction of a municipal water line to distribute potable water from the City of Byron municipal water supply to the residents at Rock River Terrace and Dirk Farm areas (Acorn and Razorville Road residents). This action along with a monitoring plan constitutes the permanent remedy to the contaminated water problem.

In September 1986, U.S. EPA issued a ROD to install carbon adsorption POE "whole house" treatment devices in the Rock River Terrace. The POE systems were either placed in an outdoor insulated shed or in the basement. The units were engineered and installed by North American Aqua, Vandalia, MI. Each system consists of a 0.5 µm (2 x 10^{-5} in) prefilter followed by two granular activated carbon (GAC) units in series. Each GAC contains 50 kg (110 lb) of Calgon carbon. Each unit is 137 cm (54 in) tall and is designed for a flow of 0.47 l/s (7.5 gpm). The system is monitored on a monthly basis before and after each carbon tank to assure the efficiency is maintained. Upon breakthrough the carbon will be replaced. To date, sampling results have shown no breakthrough of the carbon.

The next Superfund site that will be discussed is the Main Street Well Field and associated actions in Elkhart, IN. See Figures 3, 4, and 5 for maps of the Superfund site investigation areas.

Elkhart is located in North Central Indiana at the confluence of the St. Joseph's and Elkhart Rivers. The population of Elkhart is approximately 65,000. The city is diversified in the manufacturing operations that the city is known for; especially pharmaceuticals, band instruments, recreational vehicles, and injection-molded plastics.

The surface geology of Elkhart consists of a typical glacial deposit created from various types of sand and gravel, forming an extensive outwash aquifer permeating up to 53 m (175 ft). An intermediate nonpermeable clay bed confines a deeper aquifer.

To date at least five separate Superfund actions have been taken in and around Elkhart, IN. One of these actions was a remedial-type response, and four of the actions were removal-type responses. Although all are referred to as the Main Street Well Field, each removal is separate and distinct from the actions at the actual North Main Street Well Field Site listed on the NPL.

The remedial action is referred to as the North Main Street Well Field. Through routine monitoring, ground water in 9 of the 17 wells at the Municipal Facility were found to be contaminated with approximately 95 µg/l TCE. The site was added to the NPL on December 1982. Through the Superfund process, U.S. EPA and Indiana Department of Environmental Management (IDEM) decided to install packed air stripping towers at the municipal city water utility to meet the drinking water standard.

In the fall of 1987, Calgon constructed the three 17-m (55-ft) air stripping towers, while the U.S. Army Corps of Engineers supervised the construction. Each concurrent flow tower is 3 m (10 ft) in diameter and contains 9 m (30 ft) of polypropylene packing media. An estimated 19 to 23 million l (5 to 6 million gal) of water are treated per day. The total cost for construction is $2.5 million. The annual O&M cost is estimated to be between $81,000 and $106,000.

In addition to the remedial action, at least four separate Superfund-related removal actions have been taken. Two of those actions have been referred to as the East Jackson area and County 1 area.

The contamination in the East Jackson area was first recognized in the fall of 1984, when a citizen sampled his well water. Results of the sample analysis exhibited levels of TCE above 200 µg/l. These levels exceeded the 10-day health advisory of 200 µg/l.

When the County Health Department was contacted with the information, it initiated an extensive sampling program. When the County Health Department confirmed widespread contamination, which was more extensive than anticipated, it contacted the U.S. EPA.

In May, 1985, U.S. EPA conducted extensive sampling of the area whereby over 500 samples were collected. The results of the sampling program showed that heavy contamination existed in the East Jackson area where over 80 wells were found to have ground water contaminated TCE in excess of 200 µg/l, and 15 of these wells contained levels of TCE above 1,500 µg/l TCE.

Representatives from the Center for Disease Control (CDC) advised U.S. EPA that contamination greater than 1,500 µg/l is unfit for bathing and other household uses because of the inhalation and absorption dangers. As an immediate short-term remedy to the contaminated water, U.S. EPA placed approximately 800 residents on bottled water for drinking purposes within 36 hours.

U.S. EPA initially decided to extend the water main to the 15 homes with the highest levels of contamination. However, the U.S.EPA decided to blanket the area with an alternative source of water, due to the severity of contamination, its extent, and direction of ground water flow. In October 1985, the U.S. EPA on scene coordinators (OSCs), Jack Barnette and Ken Theisen, were charged with the responsibility of coordinating the installation of additional footage of water main. In total, approximately 4,420 m (14,500 ft) of water mains were provided to the town, and installed at some 300 homes and businesses. Construction of the first group of mains, amounting to 884 m (2,900 ft), was completed on December 1985. However, since various delays were experienced, the project was not completed until September 1986.

In addition, at 11 homes where the water exhibited minor contamination, POU devices were installed

Figure 3. Superfund site, Elkhart, IN.

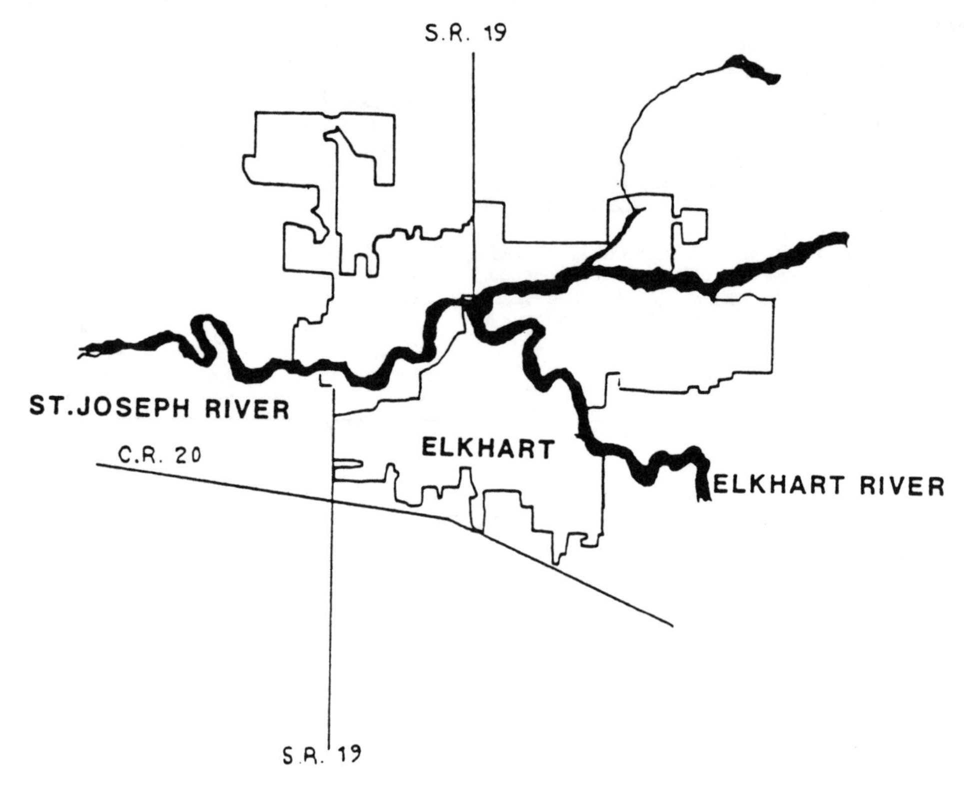

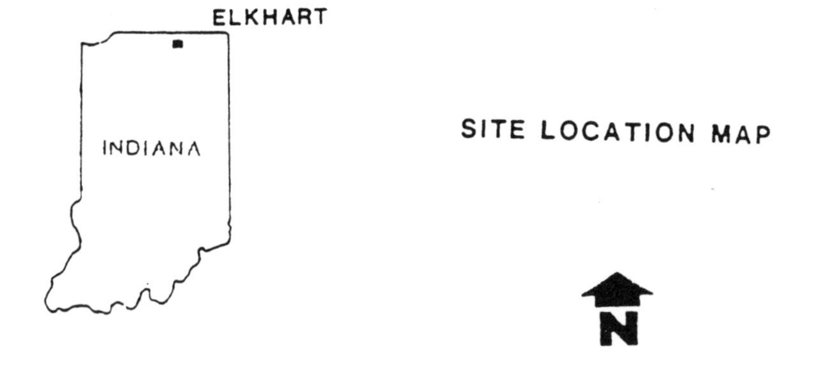

Figure 4. Superfund investigation area, Elkhart, IN.

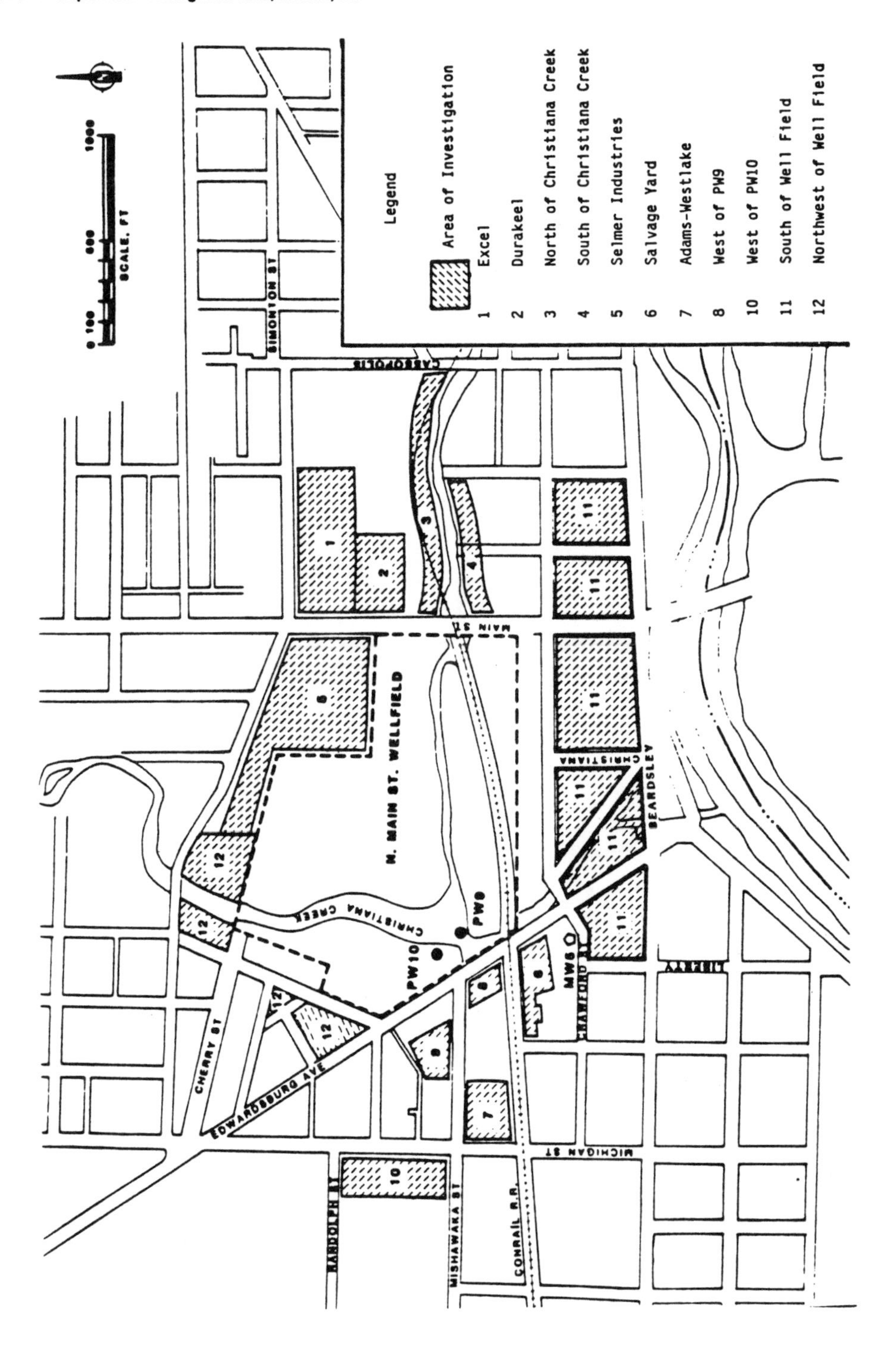

Figure 5. Investigation area, Elkhart, IN.

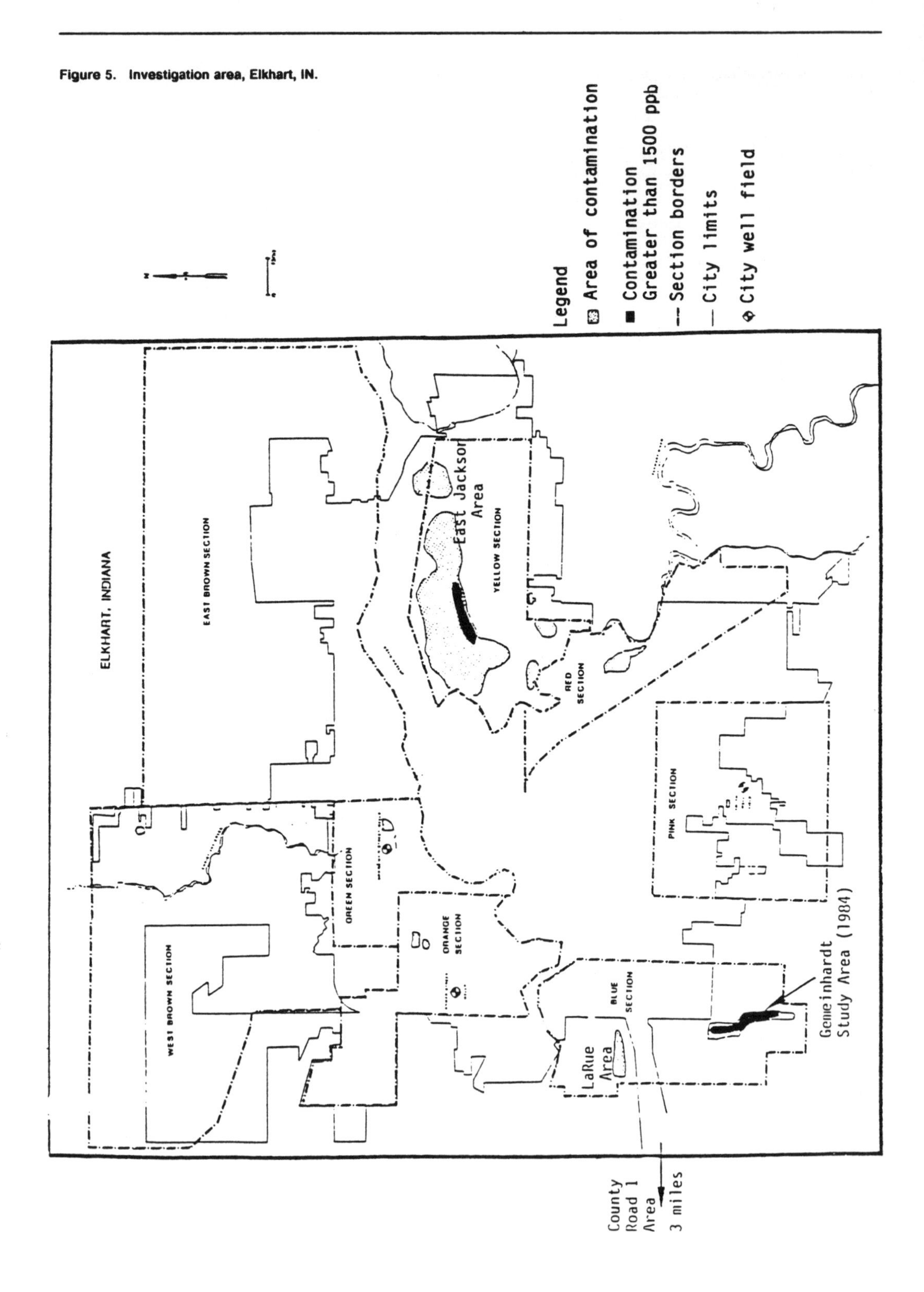

because these homes were not adjacent to the water main.

Since the area of contamination extends over 3.2 km (2 mi), it is believed that the contamination originated from more than one source. It is suspected that one of the potentially responsible parties (PRP) is the former Accra-Pac site where 13 underground storage tanks were removed by U.S. EPA.

The other removal action near the western end of Elkhart that will be discussed is the County Road 1 area in Osceola.

In June 1986, a resident analyzed the water and found contamination of TCE at approximately 800 µg/l, exceeding the 10-day health advisory of 200 µg/l, and contamination of 480 µg/l of carbon tetrachloride, exceeding the 10-day health advisory of 20 µg/l.

During July and August through extensive sampling, U.S. EPA tracked the contaminant plume from the County 1 area to its discharge into the St. Joseph River. The sampling also indicated levels of TCE as high as 5,000 µg/l, and levels of carbon tetrachloride as high as 7,500 µg/l. As an immediate response to the contamination in this area, U.S EPA provided bottled water to all affected residents, and advised them of the risks. In addition, the worst homes were advised not to use their water for any reason.

U.S. EPA decided that it would take too long to hook these residents to the municipal water supply and therefore decided to equip the affected residents with POE treatment devices. In the County 1 Area, U.S. EPA installed whole house activated carbon treatment devices onto 54 homes, and installed POU treatment devices for 22 homes located outside the contaminant plume. The last filter was installed on April 3, 1987. All units were engineered and installed by North American Aqua. U.S. EPA also initiated an extensive monitoring program, whereby the POU treatment devices are periodically sampled and analysis of the water is performed to check the efficiency of the treatment devices. IDEM has pledged to sample the affected homes and advise the homeowners.

It is suspected that the Conrail Yard south of the County Road 1 area is a potentially responsible party for the source of contamination. The site is being scored for the NPL.

Following are some of the future developments in the area of POU/POE treatment devices used at Superfund sites.

U.S. EPA has funded a pilot project in the East Jackson area of Elkhart, IN. As a prototype, two of the homes with contaminated water were equipped with a packed air stripper, along with two GAC units in series. The air stripper was placed in the basement with the GAC units and it is vented outside. The air stripper is manufactured by Tykk. The unit has a 40:1 air to water ratio, and operates at a rate of 0.32 l/s (5 gpm). The air stripper is packed with 2.5 cm (1 in) diameter polypropylene cylinders. The cost of the POE GAC unit with air stripper is approximately $4,000. North American Aqua recommends flushing the system anytime when water has had to stand for more than a day without use. U.S. EPA is considering putting on an ultraviolet light for potential bacteria problems because it is suspected that a potential exists for buildup on the media. Since U.S. EPA is still gathering data on the system, no formal results are yet available.

U.S. EPA Hazardous Waste Environmental Research Laboratory (HWERL) has funded a 12-month study for the purpose of producing a Guidance Manual for OSCs to use POE treatment devices. The study is expected to consist of first collecting POU/POE data from various Superfund sites, and following up with a pilot study to fill in the missing information. Frank Bell and James Goodrich are the U.S. EPA technical advisors.

Another project in the works is the National Register for Drinking Water Treatment Technology (The Register). Working on this project are James Goodrich of U.S. EPA's Drinking Water Research Division, along with Harry VonHuben and Sheri Bianchin of U.S. EPA Region V, Drinking Water Section. The purpose of the project is to create a National Register or data base of nontraditional, innovative water treatment systems which are being utilized to treat contaminated drinking water supplies. The majority of these technologies are being used for water supplies that have been adversely affected by Superfund sites. Additionally, Region V will serve as the test region for the Register's development.

The objective of the project is to systematically collect, organize and disseminate information on those treatment technologies which have been already implemented on either a pilot or full-scale basis at the affected supplies. The data base information will include: manufacturers and designers of the units; the design specifications; a comparison of the design performance versus the actual performance, capital and operation and maintenance costs; and a follow-up on operation problems and benefits, among other items. Presently data collection questionaires are being developed. The project then entails soliciting data through the U.S. EPA Regional offices. The data will be stored on a personal computer (PC) at the U.S. EPA Research Division, Cincinnati. Additionally, each region will be provided with a diskette for its own use and the National Technical Information Service (NTIS) will have a paper copy of information. We anticipate that the summary report and the PC diskette will be available by the close of 1988 for Region V.

In conclusion, the Superfund program is a significant part of our national response to one of the major environmental challenges of the decade. The program is a coordinated effort of Federal, state and local governments, private industry and citizens. However, since the problems are widespread and each is unique, new and existing technology is needed to remediate the hazards. In a field where no clear cut answers exist, the use of POU/POE, an available technology, has gained more acceptance in the usage and remediation of hazards where drinking water is a problem.

NEW DEVELOPMENTS IN POINT-OF-USE/POINT-OF-ENTRY DRINKING WATER TREATMENT

Gary L. Hatch
Ametek, Inc.
Sheboygan, WI 53081

INTRODUCTION

Over the last several years the U.S. EPA has promulgated National Primary Drinking Water Regulations for a number of specific drinking water contaminants, the latest of which is a group of volatile organic contaminants (VOCs) (1). These regulations, as established by the Safe Drinking Water Act, have provided impetus for the development and application of point-of-use (POU) and point-of-entry (POE) drinking water treatment technology to help solve small community water systems' and individual home owner's water contamination problems. The above referenced action by the EPA actually allows the use of POE technologies (with conditions) as acceptable means of compliance with the VOC regulations. POU technologies may also be used, but under more restricted conditions.

BACKGROUND

New developments in POU/POE technology primarily have been for contaminant-specific applications. For example, radon has come to national attention recently as being a serious health threat to private well owners. This problem is now being solved by new POE systems that employ old or well-known technologies, such as the use of activated carbon and aeration. Lowry (2) has conducted many studies to demonstrate the effectiveness of granular activated carbon systems and aeration systems for removing high levels of radon from drinking water.

Other new developments have been in the area of water disinfection. In the last few years, the well-known technologies of ultraviolet light and ozonation have been designed into packaged systems hardware for POU/POE application (3-6). Of these, the new POE ozonation systems that are becoming commercially available may be the most promising from the standpoint of overall treatment capabilities. POE ozonation systems combine the technologies of filtration and adsorption to provide for effective removal of some inorganic and organic contaminants while at the same time providing for disinfection against bacteria, viruses, and protozoan cysts. Rice (4-6) has given a detailed description of the use of ozone and ozonation systems for POU/POE applications.

During the last 5 to 10 years, the technology of home reverse osmosis drinking water treatment systems has advanced to become a viable treatment method for reducing certain health-threatening inorganic contaminants. These systems are most applicable for reducing nitrate, fluoride, and arsenic. POU reverse osmosis systems have been tested successfully for the reduction of fluoride and arsenic in a small New Mexico community (7). Proper system and membrane application is very important because various membranes (e.g., cellulose acetate versus thin-film composite) have different performance characteristics (8), especially where nitrate reduction is concerned.

In the last 10 to 20 years, a relatively new approach to using halogens for water disinfection has been developed. This technology is based on the combining of the halogen (in the form of a polyhalide) with an ion exchange resin. The remainder of this paper will provide a detailed look at the history, development, and uses of the halogenated resins for POU/POE water disinfection.

HALOGENATED RESINS FOR WATER DISINFECTION

HISTORY OF DEVELOPMENT

Initial work (9) in the field of halogenated resins began in 1957 when simple ion association experiments were conducted by adding bromine and iodine to anion exchange resins. This work revealed that these resins have an unusually high affinity for halogens, especially for iodine. Approximately 10 years later, the first halogenated resins (10) were developed to control microorganisms in water. Since then a number of improvements and modifications have been made in the formulations (11-17) of brominated and iodinated resins to enhance their chemical characteristics and their anti-microbiological performance.

The initial and primary application of the original resin systems was to use the resin as a way of metering the halogen (e.g., bromine or iodine) into a stream or body of water. The halogen, then, was still used in the conventional way (as a residual) for controlling microorganisms.

A major discovery indicating the potential for using halogenated resins in point-of-use water disinfection was made by Mills (11) in 1969. This discovery was that when microbiologically contaminated water is passed through a bed of the halogenated resin, an instantaneous and complete kill can be achieved. Therefore, a reservoir for holding the water and the required contact time needed for conventional residual disinfection action is not necessary when using a properly designed halogenated resin water disinfection system. The superfluous halogen residual released by the resin then can be removed or deactivated if desired, immediately upon emerging from the bed.

Later discoveries by Lambert and Fina (18) have not only helped to explain the unique mechanism that provides for the instantaneous disinfection, but have shown also that halogen release by the resin is not necessary to achieve the instantaneous microorganism kill. Their studies with an insoluble triiodide resin have shown that kill is achieved through physical contact of the organism with the resin beads, and that the disinfecting quantities of halogen are produced virtually on demand.

HALOGENS

The halogens are a group of chemical elements comprised of fluorine, chlorine, bromine, iodine, and astatine. Of these, only chlorine, bromine, and iodine are used in water disinfection, and only bromine and iodine are capable of forming the polyhalides, which bind to the anion exchange resins. Table 1 illustrates the polyhalides formed from bromine and iodine and how an aqueous triiodide solution is made. The higher polyiodides (I_5^- and I_7^-) are made simply by adding the correct stoichiometric amount of iodine to the sodium iodine along with a very critical and minimal amount of water).

Table 1. Halogens Most Capable of Forming Polyhalides

• Bromine	Br_3^-, Br_5^-, Br_7^-
• Iodine	I_3^-, I_5^-, I_7^-

Example:

$$\underset{\text{Iodine}}{I_2} + \underset{\text{Sodium Iodide}}{NaI} \xrightarrow{H_2O} \underset{\text{Water Solution of Sodium Triiodide}}{Na^+ + I_3^-}$$

The corresponding polyhalide resins are made by addition of the aqueous polyhalide solution to the anion exchange resin (usually in chloride form) under very carefully controlled conditions.

RESINS

In general, ion exchange resins consist of two main types - cation exchange resins (those that exchange positively charged ions, such as calcium [Ca^{+2}] for sodium [Na^+] in the water softening process, as shown in Figure 1); and anion exchange resins (those that exchange negatively charged ions, such as triiodide [I_3^-], for chloride [Cl^-], as shown in Figure 2). These resins are usually made from the polystyrene polymer backbone and differ only by their specific functional groups.

The cation exchange resin contains the negatively charged sulfonic acid functional group: $R\text{-}SO_3^-$, where R is the polystyrene backbone. These negatively charged functional groups attract and hold on to the positively charged cations. Depending on their relative concentrations and relative affinities for the sulfonic acid functional site, different cations can exchange with others, as depicted in Figure 1.

The anion exchange resins used for making the halogenated resins also usually have the polystyrene backbone, but have the positively charged quarternary ammonium functional group: $R\text{-}CH_2N^+(CH_3)_3$, where R is, again, the polystyrene backbone. Here, the positively charged functional site holds the negatively charged anion. Anion exchange occurs when the relative affinity for one anion wins out over another, such as in Figure 2 where the resin's affinity for triiodide is much greater than for chloride.

When this triiodide resin is made properly, virtually nondetectable levels of iodine are found in the post-column water effluent. Furthermore, when this resin is challenged with water high in salt content, such as with chloride or sulfate, no triiodide exchange occurs and only low levels of iodide ion (I^-) are found in the effluent (13). Table 2 shows how effective this resin is against five different kinds of bacteria and the *polyoma* virus (13,19). Other halogenated resins also demonstrate highly effective anti-microbial action as depicted in Table 3 (11,14,17).

Another type of resin, polyvinylpyridine, has been used to make halogenated resins (12,17). In this resin, the functional group attached to the vinyl polymer backbone is the pyridine molecule (see Figure 3). The unique feature of this resin is that the neutral (no ionic charge) functional group has a high affinity for the free halogens, specifically iodine and bromine. Therefore, when making halogenated resins with polyvinylpyridine, the free halogen need only be used. Use of the halide salt (Br^- or I^-) to make the negatively charged polyhalide is not necessary. The halogenated polyvinylpyridine resins exhibit anti-microbial action similar to the halogenated anion exchange resins (see Table 3).

LIMITATIONS

Many of the variables that adversely affect the conventional water disinfection methods also affect

Figure 1. Softening process with sodium-form cation exhange resin.

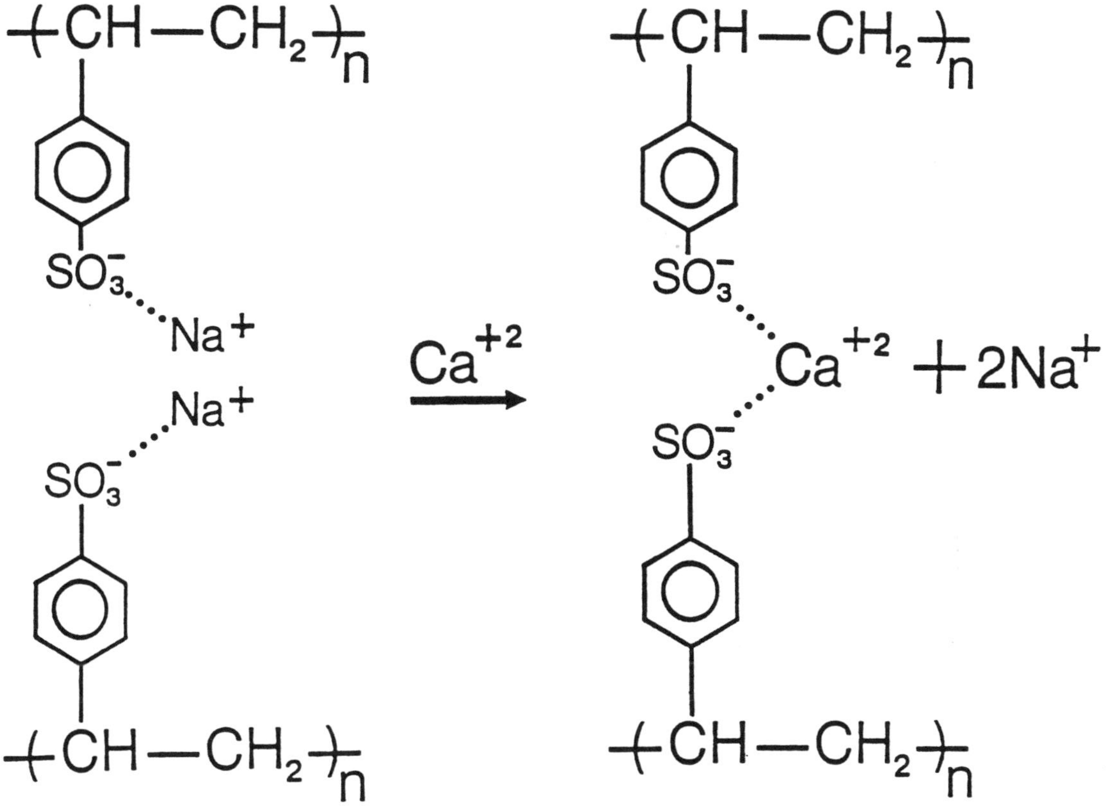

$$2\,[-(CH-CH_2)_n-]\text{-}C_6H_4\text{-}SO_3^- \cdots Na^+ \xrightarrow{Ca^{+2}} [-(CH-CH_2)_n-]\text{-}C_6H_4\text{-}SO_3^- \cdots Ca^{+2} \cdots {}^-O_3S\text{-}C_6H_4\text{-}[-(CH-CH_2)_n-] + 2Na^+$$

Figure 2. Anion exchange resins used in making halogenated resins.

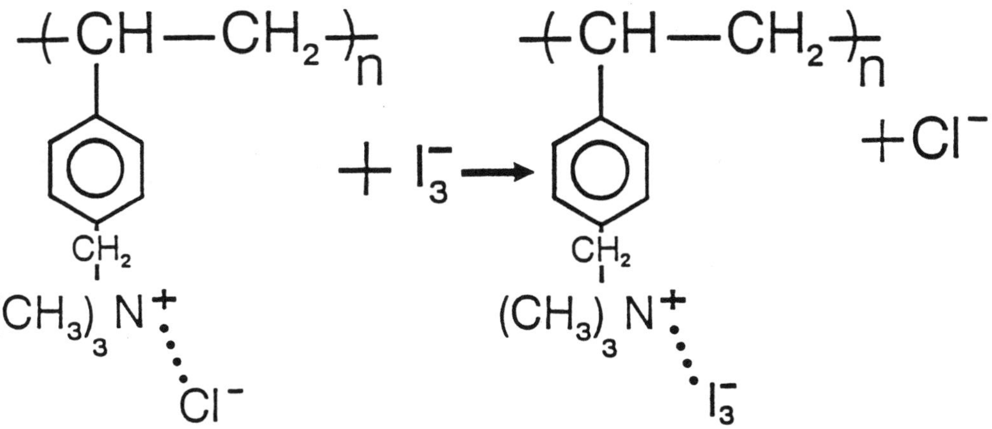

$$[-(CH-CH_2)_n-]\text{-}C_6H_4\text{-}CH_2\text{-}(CH_3)_3N^+ \cdots Cl^- + I_3^- \rightarrow [-(CH-CH_2)_n-]\text{-}C_6H_4\text{-}CH_2\text{-}(CH_3)_3N^+ \cdots I_3^- + Cl^-$$

U.S. Patent 3,817,860
Lambert & Fina

Table 2. Anti-Bacterial Efficiency of the Trioidide Resin

	Organisms per ml	
Organism	Feed	Effluent
*Salmonella typhimurium**	1×10^5	0
*Escherichia coli**	3×10^5	0
*Pseudomonas aeruginosa**	1.3×10^5	0
*Staphylococcus aureus**	1.8×10^4	0
*Streptococcus faecalis**	1.1×10^4	0
*Polyoma virus***	1.2×10^6	0

* Data from reference 13. 6.5 ml of resin at 20 ml/min flow rate.
** Data from reference 19. 30 ml of resin at 30 ml/min flow rate.

Table 3. Anti-Bacterial Efficiency of Various Halogenated Resins

	E. coli per ml		Halogen
Resin	Feed	Eff.	Residual in Effluent
Polybromide, 5% Br_2 as Br_3^-*	1×10^6	0	2 mg/l as Br_2
Mixed-form, polyiodide**	3×10^5	0	1 mg/l as I_2
Mixed-form, Iodine/Bromine-polyvinylpyridine***	9×10^5	0	10 mg/l as I_2

* Data from reference 11. 50 ml of resin at 57 ml/min.
** Data from reference 14. 50 ml of resin at 185 ml/min.
*** Data from reference 17. 50 ml of resin at 140 ml/min.

the disinfection process of the halogenated resins. Table 4 lists these major limitations as well as some other concerns of halogenated resins that may limit their uses and applications.

In a conventional disinfection process, such as a municipal treatment plant where chlorination is used, the water usually is subjected to a series of pretreatment and in some cases, post-treatment processes. Many water treatment plants pretreat the raw water by using methods such as flocculation, sedimentation, and/or filtration. Where necessary, pH adjustment can also be done. These pretreatment steps are necessary to insure that the optimum disinfection conditions are met. If low temperature or halogen demand dictates, higher levels of disinfectant can be added.

For halogenated resins, the water entering the bed also must be of a reasonable quality such that the disinfection process is not jeopardized. Therefore, a preapplication water analysis must always be conducted so that adequate pretreatment needs can be determined.

pH

If high pH is encountered (above 9), a preacidification step must be incorporated into the treatment system to lower the pH to betwen 7 and 8. Figure 4 shows the effects of high pH on the triiodide resin (20). At above pH 9, iodinated resins begin to release high levels of iodine that can diminish the kill efficiency and life of the resin bed, as well as stress and shorten the life of an iodine scavenger system. Lowering the pH to less than 5 is not recommended because of evidence that at low pH virucidal activity of the halogenated resins is diminished (19).

HIGH TDS AND HALOGEN DEMAND

The potential problem of extremely high TDS (1,000 to 15,000 mg/l or greater) could be addressed by pretreatment with demineralizing resins or reverse osmosis. These high levels of TDS could promote additional ion exchange of halide ions and halogen which, again, would result in lowering the kill efficiency and life of the resin. Similar pretreatment measures can be taken against inorganic contaminants that create halogen demand (e.g., sulfide or sulfite).

TEMPERATURE

Low operating temperature can reduce the antimicrobial efficiency of the halogenated resins (21). However, this is presumed not to be as critical as in the conventional disinfection process where the "concentration X time" constant must be maintained to assure adequate disinfection (22). The residence time within the resin bed can be increased by enlarging the bed size or lowering the flow rate to assure disinfection. Ideally, any halogenated resin disinfection system should be designed to operate at the anticipated minimum temperature. Unusually high operating temperatures such as 90 to 100°F (32 to 38°C), which may be encountered in the tropics, would most likely enhance the disinfecting action, but cause a slight increased release of halogen.

RESIN FOULING

Since the kill mechanism relies on physical contact (or very near contact) of the microorganism with the resin beads or particles, adequate protection against resin fouling or coating of the beads is a must. The effects of resin fouling are evident even in a simple softening or demineralizing process. Resin fouling or coating by iron floc or organic material, will prevent the dissolved ions from contacting or entering the resin beads, and thereby preclude the softening or demineralizing action. This same phenomenon would likewise preclude the disinfecting action of the halogenated resins. Obviously, the potential for resin fouling is an extremely critical factor that must not be overlooked when considering halogenated resins for water disinfection.

PROTOZOAN CYSTS

Any water disinfection process must be totally and reliably effective against all disease-causing organisms. A major limitation of some halogenated resins is that they are not effective against certain types of protozoan cysts, specifically *Giardia lamblia*.

Figure 3. Most recently developed halogenated resins use neutral (non-charged) resin.

$-(CH-CH_2)_n-$ + Br_2 + I_2 → $-(CH-CH_2)_n-$ ··· I_2/Br_2

Polyvinylpyridine

U.S. Patent 4,594,392
Hatch

Table 4. Limitations of Halogenated Resins

pH 9 (max.)	Halogen demand
High TDS (approx. 1,500 mg/l)	Resistance of *Giardia* or amoebic cysts
Temperature	Physiological concerns of iodine and bromine
Resin fouling - iron, TOC, turbidity	Monitoring

Tests on the triiodide (I_3^-) and penta-iodide (I_5^-) resins have shown that only the pentaiodide resin is effective against *Giardia* (23). Fortunately, since these organisms are relatively large (typically 7 to 15 µm [0.0003 to 0.0006 in]), they can be physically removed by adequate pre or post-filtration thereby precluding total reliance on the halogenated resin for a cyst kill.

PHYSIOLOGICAL CONCERNS

Since iodine does have an effect on thyroid metabolism, it is not recommended for long-term continuous use (24). Therefore, the best application of halogenated resins containing iodine would seem to be in portable emergency devices where use is short-term or intermittent. Several such devices are currently on the market.

The U.S. EPA does indicate that iodine could be used for long-term continuous application if an adequate post-treatment scavenger system were employed to remove any iodine species from the water (24). Physiological concerns with bromine are less than those with iodine, but some limitations on bromine levels in drinking water are recommended (25).

MONITORING

Any water treatment device or process that has a health effects claim should have a reliable and convenient way of letting the user know when exhaustion occurs or servicing is required. This is an obvious understatement when dealing with microbiological purification of water. One way to accomplish performance indication or lack thereof is to incorporate an automatic shut-off or remote sensing metering device into the system. The National Sanitation Foundation Standard No. 53 (26) lists other alternatives.

Figure 4. Effect of pH on iodine release from triiodide resin.

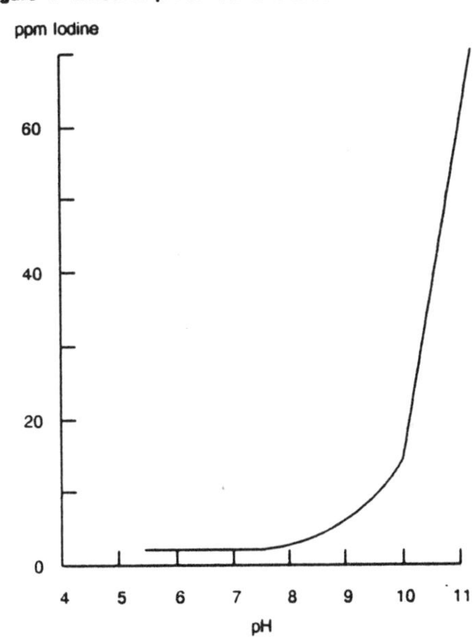

SYSTEM APPLICATIONS

The extremely efficient, once-through disinfecting action of the halogenated resins make them ideal candidates for use in small-scale point-of-use water disinfection systems. Other than previously mentioned portable emergency water disinfection devices, numerous single-tap applications are possible. Systems or units with replaceable cartridge-like containers of the resin could be used where drinking water is supplied for transient populations, such as for water fountains in parks and other entertainment facilities, or public buildings.

Multi-tap or high volume drinking water needs could be met by employing the resin in multiple-cartridge housings or in a single tank-type container such as a softener resin tank. As previously indicated for long-term use, adequate post-treatment measures must be taken to remove or reduce the halogens to an acceptable level.

Of course, as cautioned before, the critical water quality parameters must be predetermined to aid in system design, and most importantly, adequate monitoring and servicing is a must.

CONCLUSIONS

Use of halogenated resins in point-of-use water disinfection can offer a unique opportunity to help solve some of the current problems associated with conventional small-scale chlorination disinfection systems. Some advantages are:

- No electrical connections (depending on monitoring requirements);
- Easy to install and service - no handling or storage of chemicals;
- Virtually no "down" time;
- Minimal space requirements;
- More efficient use of disinfectant; and
- Potential to integrate well with other point-of-use technology.

However, even though these resins have remarkable disinfecting capabilities, proper design, thorough testing, and prudent application are absolutely necessary to assure maximum reliability of systems in which they are used. Currently available point-of-use water treatment technology can be properly combined with halogenated resins to offer another alternative for providing microbiologically safe drinking water.

All areas of POU/POE water treatment technology are advancing rapidly. Much of this technology is based on newly engineered systems, which by redesigning existing or well known technologies, can solve almost any water contamination problem. As revealed in the newly passed Safe Drinking Water Act, the U.S. EPA now recognizes that fact. The accompanying technologies of monitoring and serviceability are also advancing rapidly. As advances continue, the POU/POE industry can meet the challenge of combining reasonable cost, reliability, monitoring, and serviceability to provide safe, reliable, and economical methods of drinking water treatment for the consumer who does not have the option of central treatment.

REFERENCES

1. U.S. Environmental Protection Agency. National primary drinking water regulations - synthetic organic chemicals; monitoring for unregulated contaminants; final rule. Federal Register 50:130, 25690, July 8, 1987.
2. Lowry, J. D. et al. Point-of-entry removal of radon from drinking water. JAWWA. 79:4, 162, 1987.
3. Foust, C. Performance and application of UV systems. Proceedings. Conference on Point-of-Use Treatment of Drinking Water. U.S. EPA/AWWA, Cincinnati, OH. October 6-8, 1987.
4. Rice, R.G. Ozone for point-of-use/point-of-entry application, Part I. Water Technology. 10:3, 22, May 1987.
5. Rice, R.G. Ozone for point-of-use/point-of-entry application, Part II. Water Technology. 10:4, 28, June 1987.
6. Rice, R.G. Ozone for point-of-use/point-of-entry application, Part III. Water Technology. 10:5, 27, August 1987.
7. Rogers, K. Community demonstration of POU systems. Proceedings, Conference on Point-of-Use Treatment of Drinking Water. U.S. EPA/AWWA, Cincinnati, OH. October 6-8, 1987.
8. Slovak, J. and Slovak, R. Developments in membrane technology. Water Technology, 10:5, 15, August 1987.
9. Aveston, J. and Everest, D.A. Chem. Ind. (London). 1238, 1957.
10. Mills, J. F. U.S. Patent 3,316,173. April 25, 1967.
11. Mills, J.F. U.S. Patent 3,462,363. August 19, 1969.

12. Katchalski, E. et al. U.S. Patent 3,565,872. February 23, 1971.

13. Lambert, J. L. and Fina, L. R. U.S. Patent 3,817,860. June 18, 1974.

14. Hatch, G. L. U.S. Patent 4,187,183. February 5, 1983.

15. Gartner, W. J. U.S. Patent 4,420,590. December 13, 1983.

16. Beauman, W. H. et al. U.S. Patent 4,594,361. June 10, 1986.

17. Hatch, G. L. U.S. Patent 4,594,392. June 10, 1986.

18. Lambert, J. L. and Fina, L. R. Proceedings, Second World Conference, International Water Resources Association, Vol. II, pp. 53-59, New Delhi, 1975.

19. Hassouna, N. Doctoral dissertation. Kansas State University, 1973.

20. Hatch, G. L. et al. Ind. Eng. Chem. Prod. Res. Dev. 19, 259, 1980.

21. Kao, I. C. et al. J. Ferment. Technol. 51, 159, 1973.

22. Regunathan, P. and Beauman, W. H. Fourth Domestic Water Quality Symposium. Water Quality Assoc. and Amer. Soc. of Agricultural Engineers, Technical Papers. p. 54, Chicago, 1985.

23. Marchin, G. L. et al., Appl. Envtl. Microbiol. 46:5, 965-9, 1983.

24. Cotruvo, J. A. Policy on iodine disinfection. Memorandum to G.A. Jones. March 3, 1982.

25. Drinking Water and Health, Vol. 3, pp. 181-187. National Academy Press, Washington, DC, 1980.

26. National Sanitation Foundation Standard No. 53. Drinking water treatment units - health effects, N.S.F., Ann Arbor, MI, June 1982.

POU/POE POINT OF VIEW

Frank A. Bell, Jr.
Criteria and Standards Division
Office of Drinking Water
U.S. Environmental Protection Agency
Cincinnati, OH 45268

INTRODUCTION

I would like to preface my remarks by saying that I will not be addressing official Environmental Protection Agency (EPA) policy, since that has already been covered by Steve Clark. My remarks will also not cover pesticide regulatory programs in EPA, since that has also been covered by another speaker.

Instead I will address two basic positive options for handling the broad field of POU/POE water treatment in terms of claims control and consumer service.

OPTION #1: A DIRECT REGULATORY PROGRAM

From my understanding, such a program would involve a verification of claims being made by the various industry segments. To be universally applicable, such a program would have to identify and circumscribe acceptable claims and then have units tested against the claims. This would be an incredibly complex operation. It also might be considered a restraint of trade. Some questions that it would present are:

- How would claims be limited?

- Would the direct regulatory program deal only with physically or chemically verifiable claims? If so, would noncovered units be exempted, such as magnetic treatment units and good-health units (ones that treat the water to improve its good-health qualities without any change in chemistry)? If such units are exempted, then a substantial area of consumer concern will remain unattended.

- Would nonhealth or esthetic treatment units be covered? If yes, then the magnitude and complexity of the program would become serious problems, since from a review of water conditioning treatment (1) some 26 possible esthetic effects/treatments emerge. If, on the other hand, esthetic treatment units are excluded, what will be done about the incidental health effects claims of the esthetic units? For example, a taste and odor activated carbon filter may also remove 10 to 20 percent of the trihalomethanes (significant, but not enough to be considered adequate from a health standpoint). If the 10 to 20 percent reduction claim is allowed, how can it be described so as to inform the consumer properly?

- The currency question will provide further enormous problems. Manufacturers are constantly changing their products. How will these changes be monitored? How will retests and quality control be verified? Who will certify treatment capabilities for the manufacturers' new products? For a government agency to attempt to regulate, monitor, and control the entire water quality industry would represent an enormous and continuing bureaucratic effort.

CONCLUSION

While a direct regulatory program has some appeal from a simplistic viewpoint, it can introduce a host of hard-to-answer questions, and in the end, provide more problems than solutions.

OPTION #2: COOPERATIVE INDUSTRY/GOVERNMENT/THIRD PARTY EFFORT

Historically, EPA has vigorously supported this second option. I have personally participated in several productive efforts that have involved many cooperative parties, including:

- The Gulf South Research Institute's landmark study of over 30 commercially available activated carbon and other units for organic chemical reduction capabilities (2). Before this study, no authentic, independent information on unit capabilities existed. Industry cooperated in protocol review and development, and in the conduct and review of the study and its results. Industry technical representatives helped us to avoid doing unwise things and in having a valid, widely accepted study result.

- Establishment of third-party standards and listing program under the National Sanitation Foundation (NSF). NSF has provided leadership to bring industry, government and other interests together for establishment of a number of drinking water treatment unit standards, with three currently in effect: Standard 42 (Esthetics), Standard 53 (Health), and Standard 58 (Reverse Osmosis), and with two other standards nearing adoption. Strengths of an NSF program include the standards consensus development process (giving all interests a chance to be heard) and its testing, monitoring, and control capabilities. NSF has a continuing presence and mechanism to keep up with new product developments and control of product quality over time.

- Water Quality Association (WQA) industry survey for use of solvents in the manufacture of water treatment units. In response to EPA's concern regarding possible solvent contamination in drinking water from some home treatment units, WQA conducted an industry survey and developed guidelines for the use of solvents (3). This survey and guideline development has raised the consciousness of the water quality industry and, I believe, minimized the inadvertent contamination of water from home treatment units.

I could enumerate other areas of industry/government cooperation to include the microbiological purifier guide standards project or the concern with contamination from ion exchange resins. However, I believe the cooperative atmosphere is well developed.

CONCLUSION

While cooperative efforts may suffer slightly by not having regulatory backing, they can sometimes be the best solution, particularly if the interests of the parties are sufficiently served by such efforts. In this situation, I believe Option #2 for cooperation is the optimum choice.

RECOMMENDATIONS

Additional steps need to be taken to strengthen consumer protection in areas of POU/POE.

- The water quality industry needs to continue and expand education and training efforts aimed at raising the knowledge and professionalism of water quality contractors, salesmen, and technicians. Eventually, the contributions of the water quality industry will be limited or expanded depending on the professionalism and credibility of its personnel.

- More attention needs to be given to the expansion and utilization of the third-party standards and certification program. Specific areas for attention are information and education programs for local and state government personnel and for consumers regarding the NSF programs, and greater utilization of the NSF listing service by water quality product manufacturers. The third-party program will not be truly effective until it is more widely recognized and utilized, particularly by local and state regulatory officials.

- Finally all Federal, state, and local officials and industry and other interested parties need to maintain open channels of communication to examine areas where problems arise. Continuing the positive patterns of the past will enable us to have a productive result in the future.

REFERENCES

1. McGown, Wes. Sensitivity: a key water conditioning skill. Water Technology. September/October 1982, pp. 2-5.

2. Bell, Frank A. et al. Studies on home water treatment systems. JAWWA. April 1984, pp. 126-130.

3. Water Quality Association. Voluntary guidelines for the use of solvents. July 1987.

AWWA VIEWPOINT ON HOME TREATMENT UNITS

Jon DeBoer
American Water Works Association
Denver, CO 80235

INTRODUCTION

In order to present the viewpoint of the American Water Works Association (AWWA) regarding point-of-use (POU) or point-of-entry (POE) treatment units, it is necessary to provide a brief overview of AWWA.

Many are familiar with the *Journal AWWA*, perhaps *Mainstream*, *OpFlow*, or *Water World News*. In many cases these publications are the only perceptions people have of AWWA. But AWWA is not just the efforts of its staff, it's the efforts of the people involved in the industry -- the engineers who design systems, the people in the field who install systems, and those who treat and analyze water.

AWWA is involved not only in drinking water, but also water for all uses. This includes not only commercial, industrial, and other city uses, but also many uses within the home such as drinking water, water used for cooking and cleaning, and water used for consumption at taps other than the kitchen sink. AWWA is a service organization, and when individuals have questions, we are available to provide answers.

The organization is made up of members divided into sections (regions of the country). Each section elects a member to the Board of Directors, the controlling body of the organization. The Board of Directors internally selects an Executive Committee, which oversees the entire operation. AWWA is divided into four councils, the Water Utility Council, which is our newest council, the Technical and Professional Council, the Standards Council, and the General Policy Council.

THE POINT-OF-USE/ POINT-OF-ENTRY ISSUE

Each of these councils is involved with POU/POE issues. The Water Utility Council is involved since this issue impacts utilities, particularly from a legislative and regulatory viewpoint. They are primarily responsible for providing input to the regulatory and legislative positions developed on both national and state levels. The General Policy Council is involved because AWWA is in the process of establishing a policy or position on POU/POE. The General Policy Council is set up only to do this: to oversee the development and planning of Association policy. The Technical and Professional Council is involved because there are technical issues. The Technical and Professional Council's scope is to deal with the technical issues in the water industry. They are not involved in the development of standards; this is reserved for the Standards Council.

This paper discusses AWWA's position and viewpoint. It is necessary to understand that AWWA has two types of statements. One is a policy statement, the other a position statement. The difference between these two is the way they are developed and used. Policy statements have been in existence for a long time; they are long-standing positions, and have been thoroughly researched. They are selected by a rigorous process, and they require Board of Directors' reaffirmation at least every five years.

Position statements, on the other hand, are a position on a specific issue. They are often developed in response to regulation or legislation. They are somewhat more fluid, and rather than Board of Directors approval, they are considered official policy once approved by the Executive Committee. However, the Board of Directors reaffirms a position statement at its next meeting, and it must be reviewed annually to ensure that it is still a current and valid position of the Association.

LEGAL REQUIREMENTS

The primary legal issue, from the AWWA viewpoint, is the Safe Drinking Water Act Amendments and the requirements the Act contains for improving water quality. The utilities and the bulk of our membership believe that all water should be treated to acceptable quality for all users and for all uses. Frequently, the home treatment industry believes that it is more appropriate to provide marginally treated water for general use and provide independent treatment of specific water streams that have higher use requirements. We have seen comments published where municipally supplied water is described as raw or untreated water. A POE device might be supplied to provide treated water for other specific uses such as cleaning and laundry. Finally, further treatment by an under-the-sink device might be used to increase the quality of drinking water. AWWA believes that all water supplied to a home should be considered drinkable.

By law, in the development of the National Primary Drinking Water Regulations, the purveyor is responsible to the tap. That doesn't mean through the meter to the connection. It means to the tap, and it means to every tap within the home. The regulations are to provide protection against possible, as well as proven, harmful agents, which expand the number of contaminants that may or probably will be regulated in the future.

In addition, EPA is required to define acceptable treatment techniques. Best available technology does not include point-of-use devices for VOCs; that question is still open for regulation in the future. In other words, each best available technology will be defined based on the contaminants that are considered.

In addition, affordability can be taken into account by the agency in defining best available technology; this is one criteria that has been tentatively defined. An increase in the annual water bill for removing a contaminant shall be limited to no more than one percent of the household income, with a total annual water bill of no more than two percent of income. That doesn't leave a great deal of money for either point-of-use or central treatment. The goal of the Safe Drinking Water Act and the objective of the AWWA is to consistently meet the needs of the public. We feel that these needs must be met for all people, not just selected populations who can afford to install a treatment system.

AWWA'S POSITION

The AWWA position on POU/POE treatment devices is: aesthetic treatment of potable water is something that AWWA has no objection to and never has had. We commend the Water Quality Association for its establishment of Voluntary Guidelines and the use of the Industry Review Panel. We encourage, not only the manufacturers, but also the utilities, to use this system to ensure that promotion of products is fair and accurate. These products must be properly advertised and must not condemn municipal water supplies. That kind of advertising does neither our industry nor the home treatment industry any good. Advertising is only one form of promotion, the other is the actual demonstration or application of a unit in the home. Frequently we hear cases where the term "snake oil salesman" might apply. These individuals do not do our industry any good, they do not do the POU/POE industry any good, and we don't think they're appropriate.

Treatment for health effects to meet current regulations is another matter. This is where AWWA has developed a draft position statement. The position statement development procedure includes review by numerous councils, divisions, and committees, based on their expertise of a specific topic. Following these reviews, the General Policy Council reviews the entire position statement and develops a final draft based on the comments that have been received by all the review bodies within the Association. In the final approval process there may be some modifications to the wording and language, but not to the specific intent of the statement. The following will describe the general content of the statement without using the specific words:

AWWA believes that POU/POE devices are not appropriate alternatives or replacements for central treatment of drinking water. Central treatment of drinking water is the alternative of choice. There are occasional situations where POU/POE treatment might be appropriate. But the condition that we believe is necessary is central control, although ownership may not be necessary. Rental/lease situations may be entirely appropriate for POU/POE treatment, but the control still has to be under the auspices of the water utility.

In addition, there has to be an effective monitoring program established. POU/POE units have to be properly applied. That means there has to be an engineering and health review of the units which includes how they are going to be installed in the home. Microbiological safety must be maintained. We have to protect all consumers at all points where they are going to consume water, both intentionally and inadvertently. We can't assume that people aren't going to come into internal contact with water simply because we didn't design it to be a point of drinking water. We believe that there should be no increase in the risk over a centrally treated supply.

A POU/POE device is installed specifically for modification of water quality in a beneficial way. However, there are risks. There are possibilities that, in addition to the beneficial modification, there is a potential for adverse modification of the water. We believe that consumers need to be educated to understand the potential for adverse modification with POU/POE devices.

This is the basic content of the AWWA draft position statement. It has been developed as a consensus of the entire industry, not restricted to utility input. There are numerous people such as academics, consultants, utilities, and manufacturers on the committees and councils, who have reviewed this position statement. It is not a frivolous matter. As stated earlier, it is in the final stages of development at this point. Once approved, it will be published in Mainstream. We hope that this discussion has brought a better understanding of the AWWA position on point-of-use devices.

POU/POE - POINT OF VIEW - ASSOCIATION OF STATE DRINKING WATER ADMINISTRATORS (ASDWA)

Barker G. Hamill
Bureau of Safe Drinking Water
New Jersey Department of Environmental Protection
Trenton, NJ 08625

Before I make my disclaimer about who I actually represent, as my function as a Chief of the Bureau of Safe Drinking Water in New Jersey, I would like to comment on the previously mentioned New Jersey study. We are still trying to deal with the ramifications of actually having two incidents within one county in New Jersey where there were installations of treatment units that did not prove satisfactory. We have not been able to get the county health officials to allow any installations of home treatment units for health-related matters since then; the consequences of what happens when things go wrong can take a long time to overcome at the local level. Now that I've said that, as an official from New Jersey, I would also like to say that to represent all 50 states is obviously an impossible task. The views that I have are mostly my own, but I have gone over them with some of the people within the association. There will be differences in how the individual states react to point-of-use treatment, so I'm talking about things in a very general sense in terms of representing the association.

I've always been impressed with EPA's research efforts, their speakers, and the type of research that they do. As a state official, I was very impressed with the technical presentation by the point-of-use treatment industry at this meeting. I have not seen too many presentations by pointof-use treatment people before, and I was very surprised by the technical expertise of the point-of-use treatment industry represented at this meeting. I think this is something that the state officials need to take back to their local people: there is good technical work being done within the point-of-use treatment industry.

The good aspect of having a large number of manufacturers and installers is that it allows for a lot of attention to local needs. A lot of the needs within the point-of-use treatment industry are of a very localized nature. There are different needs for example, in Arizona, New Jersey, or in Florida, and it's important to have different manufacturers and installers serve different needs. It was good to see how many actual field experiences have been a success, and to have everyone learn from these field experiences. I also think the beginning of concensus industry standards is a good thing, whether it's for aesthetic effects and performance characteristics done by the Water Quality Association (WQA) or whether it's the health related work of the National Sanitation Foundation (NSF). I think its a good beginning.

There is however a need for what I consider the transfer of technology. We need to transfer information about the industry from the national level to the local and individual level. This technology transfer will determine how much regulation gets involved in this entire process. We are going to need to get technical information to the local health officials because if good things don't happen at the local level, the response is legislation. To accomplish that, I think we need to continue to build on some sort of design and performance guidelines, both within the WQA and the NSF or any third-party certification program that emerges. I think it is a very good idea that we consider things like the use of a surrogate parameter like chloroform to see how design standards are made useful, and to look at the application of performance standards. We will need to apply performance standards and to realize that there still will be various state standards, not only in what will be applied, but also variations in what the maximum contaminant levels are. I think one of the first problems that the industry is going to have to deal with is the various states' standards. For example, from my own experience, for a chemical like 1,1,1-trichloroethane we are going to wind up a year from now with three standards in use. There's going to be the national standard, 200 ppb, which EPA has regulated, a standard in New Jersey that's going to be 26 ppb; New York is well on its way to having a standard of 5 ppb. The different local and state health officials in those areas will need to have enough performance information about GAC units to be able to make good decisions on what size unit to put in for what size household. It's going to be important for the manufacturers in the industry to develop information that can be easily understood and transferred to the local level for these types of differences in regulatory efforts.

I want to discuss the down side of having a large number of manufacturers -- it makes it harder to get all this done on a consensus basis. It is tougher to transfer the information, and it also means that problems associated with the introduction of new technologies are blown out of proportion. Something that I see in the future in terms of New Jersey, is home treatment or point-of-use units for volatile organic chemicals, both within private wells and what I consider the new EPA classification of water systems that are non-transient. It's an area that hasn't been talked about a lot during this conference, but I think that these types of water systems coming under new federal regulation have more monitoring, are going to have the ability to meet these new standards, and are going to cause both the states and industry to address those problems. I think radon will also continue to be an area where home treatment units are going to be an integral part of any state strategy. I think there are some nitrate problems in New Jersey, and I assume there are other states that are going to have those problems as well. I also think lead is an issue that was not addressed here. Depending upon where EPA goes with its regulations, there could be a need to look very closely at lead standards and home treatment units, both in terms of private individual wells and in terms of public water supplies for those spots that can't be treated otherwise.

When I look down the road a little bit, I see four different options. We could have a huge federal program or a partial federal program which has registration, certification, design, and performance standards. This will go a long way to do certain things in terms of assuring public health, but it also stifles new innovative designs and puts a financial burden on the whole process.

We can have individual state programs with individual sets of state regulations and state registration and certification programs in some states; in other states there won't be many programs. I'm not sure that people will like that. I know that the person from California didn't say that their program started from the regulatory effort of the state officials. It was started by the legislature when they had problems in that state. I don't know of many state regulatory programs that would initiate a program themselves. We don't necessarily look for more work to do unless we feel that there's a huge public health gain. We're not sure in these instances whether there is a gain or whether there isn't.

Third is the direction we seem to be going, which is consensus guidelines, third-party certification, and local and state application of these guidelines and certifications. This has a lot of benefits to it, both from the industry standpoint and the state standpoint. But it's got to do the job, and it's got to be perceived by the public, the federal agencies, and the legislatures of doing the job of protecting public health. And if it doesn't do that job, then we're going to wind up with a bigger regulatory effort since one of the overriding principles is that the level of regulatory effort corresponds to the level of public health concern. If you have more concern that things aren't working right, the chances of legislation and regulation are going to increase.

The fourth option would sort of go backwards -- not to do anything at the federal and state level and let all the local people take care of the problems. It's a local individual choice. My general belief is that we have the opportunity to make a good start, and we've had a good start at consensus guidelines and third-party certifications. We are going to have to continue that effort and make it work. The less that we make it work, the more likely we will get federal regulations or individual certification programs and legislation for each state.

POU/POE: AN INDUSTRY PERSPECTIVE

Donna Cirolia
Water Quality Association
Washington, DC 20005

In order for the audience to better understand where the point-of-use/point-of-entry (POU/POE) water quality improvement industry is today, and where it's going, it is important to gain a historical perspective on where the industry has been. Back in 1976, the Water Quality Association (WQA) had only 877 members. Today we have over 2,000 members. At our convention, which we hold every year in March, we used to have nearly 100 exhibitors. In the last five years, the number of exhibitors has increased to over 200. Coupled with this expansion, the direction of the industry has changed from products dealing solely with aesthetics, to products reducing contaminants. However, this transition into the health arena has led to a number of new issues and concerns.

WQA and others have developed programs to address these areas, including product validation and personnel certification, which is a WQA program designed to educate our dealers who serve the consumer.

I certainly agree with some of the other panelists that our industry needs to do more in terms of credibility and accountability for both products and personnel. However, we have made great strides. Today, I may be preaching to the choir, but it's important that the industry members in the audience encourage your distributors and dealers to become better educated.

We have some fabulous companies that provide quality products and quality service. I find it discouraging to see that some people outside of our industry unfairly judge the entire industry by the minority that use misleading sales tactics. WQA recognized the need to address the issue of misleading advertising that occurs in our industry, as in others, and developed the Water Quality Improvement Industry Voluntary Product Promotion Guidelines. This program is a positive step in the right direction.

The industry's advertising, products, and services are constantly improving. Our efforts have been met with greater recognition by EPA and the states, and this seminar proves that.

Where our industry needs to do more is in our relationship with the water utilities. However, they also need to do a bit more. I was disappointed that very few water utility people attended this conference, particularly since AWWA was a co-sponsor. It would be far better for both industries not to dwell on the negative aspects, but rather change with the times and adjust their attitudes so that we can both be partners in providing quality water. We have products that can solve health problems, and also improve the aesthetic quality of centrally treated water, if the consumer so chooses. This is based on our system of free enterprise, where individuals have the right to choose products of their liking. Utilities should not feel that the mere existence of the POU/POE industry undermines the quality of water which they serve to their customers. If a consumer does not like the taste, odor, or color of their water, they have every right to install a POU device.

Presently, a few states are considering their own mandated product validation standard to provide consumer protection. If WQA really felt that this was the answer on how to provide consumer protection, we would be out there promoting this type of legislation. However, we really don't feel that product validation is the solution. In every state there are consumer fraud laws; the problem is in the lack of enforcement. We encourage attorneys general and county consumer affairs bureaus to enforce their consumer fraud laws against those companies that use misleading sales tactics. Trying to regulate the industry in terms of consumer protection through product validation is not the answer. This will just end up increasing the cost of the products and services to the consumer. The voluntary standards that exist today are in a constant state of flux, being periodically revised to meet the demands placed on the products in the marketplace. Also, it takes a long time to both develop and revise consensus standards. Realistically, this type of program would be very costly for the states to implement and enforce. I would rather see the marketplace driving our members to get their products validated, than the regulator saying, "Oh, I think we need to protect the consumer via product validation." WQA urges the states to use their existing consumer fraud laws, and if they're not strong enough, then enhance them. This is a more realistic and productive way to provide consumer protection.

Finally, the industry is at the crossroads of trying to meet the needs of many small systems and nontransient, noncommunity systems that are facing the same heavy regulatory burden as other public water systems. We're not saying that point-of-use and point-of-entry are the only answers for these systems, but they should be considered as options. Regionalization may be cost prohibitive if the next system is many miles away. These systems are going to be forced to look at new, innovative solutions including point-of-use and point-of-entry. The industry has made great technological strides concerning the monitoring and maintenance of point-of-use/point-of-entry units. We do not have all the answers but we would certainly look forward to working with any state or local government agency that is interested in the POU/POE option.

In conclusion, WQA realizes that our industry needs to do more in terms of credibility and accountability. However, I think this also has to be met by a greater willingness, particularly by the water utilities, to realize that central treatment is not the only answer. We're really partners in providing quality water to the consumer. That's who we're trying to please. I just hope that we'll have the opportunity to have more conferences such as this one so that all the interested parties including Federal, state, and local regulators; water utilities; and POU/POE manufacturers and dealers can maintain a continuous dialogue on these important and critical issues. We are all partners in providing and assuring safe and aesthetically pleasing water to the consuming public.

POINT-OF-USE TREATMENT OF DRINKING WATER: COMMENTS

Sue Lofgren
Tempe, AZ 85282

In three years on the National Drinking Water Advisory Council, as the EPA struggled to implement the Safe Drinking Water Act, I could see that, particularly as we dealt with public notification, people were confused and afraid. They would rush out to do one of two things: buy bottled water or buy a home treatment device. It's a natural reaction. They didn't wait for the public utility to come up with a new treatment for the contaminant they thought was there. As a result, I kept saying to EPA, "This law is going to give those two industries a license to steal." At that time there was nothing to regulate those industries. Therefore, EPA had a responsibility to do something to help guide the public on as to the degree of effectiveness of these devices.

There were two items I bugged EPA about: point-of-use and ground water. At that time, nobody thought ground water was a problem. In the 10 years since that time, I've seen real progress made in both areas. The industry has made an immense drive forward in terms of trying to police its own, to make sure that the devices on the market are effective, that the advertising is true, and that manufacturers aren't resorting to scare tactics.

However, we still have a long way to go. What we have been talking about today, and what I think the New Jersey study points out, is that what is still needed is something for the consumer. I'm speaking from the perspective of the public -- the average person. Where does an individual go to get something to take care of his concern? If it isn't odor or taste, smell, or sight -- if it's health concerns, how does he know what to get? First of all, he doesn't really know what his own health concern is. What is the contaminant in the water? What device does he need to remove that particular contaminant? People call me in Arizona and think that I am the person that knows something about water since I seem to sit on everybody's committee. But, the state also gets called, and they don't know how to respond. So there is a real need for something -- somewhere that the public can go to get these kinds of questions answered.

Perhaps the validation of every manufactured device is the answer in some form. You can't rely on the consumer report -- that doesn't give you the kind of information you need, because there's a diversity of contaminants out there. So, I really think EPA, the states, and industry have got to come to grips with this. I would suggest that states like New Jersey, California, and some of the others who have done some exploring along the lines of validating or testing, get together and provide a central repository. EPA or the Association of State Drinking Water Administrators might be the ones who deal with this.

Once an individual knows what to buy, he doesn't know how to maintain it. There are usually instructions, but those instructions get lost. How many people keep all the instructions that come with anything they buy that is mechanical? There are filters and cartridges that have to be changed. When do you change them? Maybe you remember the first time, but then maybe you don't. Assuming you even remember, where do you go to buy that filter, that cartridge?

I'm speaking from personal knowledge. My husband went out and bought a faucet treatment device while I was on the council, and I said "What are you getting this for?" He said, "Well, in Arizona they dry up the canals once a year, and when they do, instead of surface water, our utility provides us with ground water which tastes bad. " So he wanted something for taste. I said, "Well, in that case, that's not so bad, but are you sure it's not one of those that leaches silver into the system, into our water?" "I don't know." Well, neither did I to be truthful. That's the sort of thing I'm talking about.

More recently my son called me from Prescott, and I live in Tempe, that's about 130 miles away. He said, "I've got a friend who needs to get a cartridge for his under-the-sink system and he can't find a dealer up here. Where does he go?" I said, "Look it up in the Yellow Pages. If you can't find it there, he'll have to write the manufacturer." These are the sort of things that people shouldn't have to deal with.

The water quality industry also deals with bottled water. One of the things that we found out on the Drinking Water Council was that the bottled water industry is regulated by FDA. EPA does not have responsibility or any enforcement ability. All they can do is set the standards for FDA to enforce. Of course, FDA had many other concerns and drinking water was at the bottom of the totem pole. When we finally got to talk to the director of FDA who was responsible for this area, we found out that, not only was it on the bottom of the totem pole, it basically never got any attention. If you import mineral water, and it says "mineral water" -- nobody looks at it. It was rather amazing to us to find out how little attention is paid to bottled water.

Let me speak from where I am in terms of a person who is dealing with public involvement programs and the needs that I can see out there. I would say that you need to involve the public more in whatever you're trying to accomplish, whether you're an agency person or industry person. You need to reach the public with the information.

The future is going to call for increased use of the point-of-use and point-of-entry devices in areas where people are scattered. In Arizona for instance, our Indian tribes, which are really scattered, will definitely be looking at this as a potential way to resolve some of their problems. I urge you as an industry to continue to look at alternative forms of treatment because very few people can handle the expense of many of the conventional treatment processes for some of these contaminants. However, there have to be ways of assuring that those kinds of treatment devices are monitored and maintained. You cannot rely on the local person to be the person to maintain the devices. You may be able to train someone, but sooner or later they're gone and the next person who takes over won't be trained, and then you may really have a very bad situation. That is going to have to be one of the key points, how to keep something that is effective, in use and properly maintained.

Finally, I'll just go back to making a pitch for efforts to increase the educational level as we deal with drinking water and as we discover new contaminants. And, as we try to explain to the public that they aren't going to die tomorrow, that these contaminants, these MCLs, are based on drinking two liters of water over a lifetime of 70 years, that some of these things are naturally occurring, and that the danger is not necessarily immediate, at the same time explain that one needs to do something about it and to do it effectively.

Other Noyes Publications

TREATMENT OF MICROBIAL CONTAMINANTS IN POTABLE WATER SUPPLIES

Technologies and Costs

by

Jerrold J. Troyan and Sigurd P. Hansen

CDC-HDR, Inc.

Pollution Technology Review No. 171

This book identifies the best technologies or other means that are generally available, taking costs into consideration, for inactivating or removing microbial contaminants from surface water and groundwater supplies of drinking water. For municipal officials, engineers, and others, the book provides a review of alternative technologies and their relative efficiency and cost. More specifically, it discusses water treatment technologies which may be used by community and noncommunity water systems in removing turbidity, *Giardia*, viruses, and bacteria from water supplies.

The USEPA does not require any system to use a particular technology to achieve compliance with proposed treatment regulations. Whatever technology is ultimately selected by a water supplier to achieve compliance with EPA requirements must be based upon a case-by-case technical evaluation of the system's entire treatment process, and an assessment of the economics involved. However, the major factors that must be considered include:

- Quality and type of raw source water
- Raw water turbidity
- Type and degree of microbial contamination
- Economies of scale and the potential economic impact on the community being served
- Treatment and waste disposal requirements

The technologies and actions available to a community searching for the most economical and effective means to comply with microbiological regulations include modification of existing treatment systems; installation of new treatment systems; selection of alternate raw water sources; regionalization; and documenting the existence of a high quality source water while implementing an effective and reliable disinfection system, combined with a thorough monitoring program, and maintaining a continuing compliance with all drinking water regulations.

A **condensed table of contents** is given below.

ISBN 0-8155-1214-7 (1989) **6"x9"** **335 pages**